Strategies for Success: Math Problem Solving, Grade 3 OT111 / 322NA ISBN-13: 978-1-60161-936-5
Cover Design: Bill Smith Group **Cover Illustration:** Valeria Petrone/Morgan Gaynin

Triumph Learning® 136 Madison Avenue, 7th Floor, New York, NY 10016

Printed in the United States of America. 10 9 8 7 6 5 4 3 2 1

Table of Contents

Problem-Solving Toolkit

How to Solve Word Problems

You follow a set of steps when you make something. When you solve word problems, you can also use a set of steps.

- ☐ **Read the Problem** Read carefully to understand the problem. Be sure you know what it is asking. Try to picture what is going on. Find out what question you need to answer.

- ☐ **Search for Information** Mark all the words and all the numbers you need to solve the problem. Study any charts, graphs, and pictures. They may have the information you need to solve the problem.

- ☐ **Decide What to Do** Think about the problem. Try to find a way to solve it. Ask yourself if you have solved one like this before. Think about all the problem-solving strategies you know. Choose one that you think will work.

- ☐ **Use Your Ideas** Start to carry out your plan. Try your strategy. Think about what you are doing. Check that you are on the right track. If not, change what you are doing. There is always something else you can try.

- ☐ **Review Your Work** Look back at the question in the problem. You may have computed, but the number you got may not be the answer to the question. Be sure you answered the question in the problem.

You can use the Problem-Solving Checklist on page 7 to make sure you have followed these important steps.

Problem-Solving Checklist

☐ Read the Problem

- ☐ Read the problem all the way through to get an idea of what is happening.
- ☐ Use context clues to help you understand unfamiliar words.

Ask Yourself

- ☐ How can I restate the problem in my own words?

☐ Search for Information

- ☐ Reread the problem carefully with a pencil in your hand. Circle the important numbers and math words.

Ask Yourself

- ☐ What do I already know?
- ☐ What do I need to find out to answer the question the problem asks?
- ☐ Does the problem have any facts or information that are not needed?
- ☐ Does the problem have any hidden information?
- ☐ Have I solved a problem like this before? If so, what did I do?

☐ Decide What to Do

- ☐ Choose a strategy that you think can help you solve the problem.
- ☐ Choose the operations you will use.

Ask Yourself

- ☐ How can I use the information I have to solve the problem?
- ☐ Will this problem take more than one step to solve?
- ☐ What steps will I use?

☐ Use Your Ideas

- ☐ Try the strategy you chose to solve the problem.
- ☐ Do the necessary steps.
- ☐ Write a complete statement of the answer.

Ask Yourself

- ☐ Do I need any tools such as a ruler or graph paper?
- ☐ Would an estimate of the answer help?
- ☐ Is my strategy working?

☐ Review Your Work

- ☐ Reread the problem.
- ☐ Check your computations, diagrams, and units.

Ask Yourself

- ☐ Is my answer reasonable? Does it make sense?
- ☐ Did I answer the question the problem asks?

Guess and Check

Guess what I found? Guess what I had for breakfast? Guess what movie I saw on Saturday? It is fun to guess. Guessing is a way to solve math problems.

Take three numbers in a row, such as 2, 3, and 4. If you add 2, 3, and 4, you get 9. What three numbers in a row add up to 24?

Start with a guess. Is the answer 5, 6, and 7?

$5 + 6 + 7 = 18$. No, that is not it. The total needs to be 24, not 18.

You can guess again. This time, try numbers greater than 5, 6, and 7.

Keep guessing until you get the answer. Try to get closer with each guess.

When you use the "Guess and Check" strategy, think about each guess and check. Use each incorrect guess to help you make your next guess.

Write a Number Sentence

Amusement parks can be a lot of fun. There is so much to do! You had $10 when you got to the park. When you leave the park, you only have $1.50. How much did you spend?

To find out, you can write a sentence to describe the problem.

The amount I have left	is	***the amount I started with***	minus	***the amount I spent.***

Then you can use the numbers you know to write a number sentence.

$$\$1.50 = \$10.00 - \square$$

Can you find the unknown amount? It is the amount that will make this number sentence true.

You can write more than one number sentence for a problem. Use any number sentence that fits the problem.

Draw a Picture

Jen crosses the finish line in fifth place. Her brother Jon is the ninth runner to cross the line. How many runners finished the race between Jen and Jon?

You can draw a picture and count the runners between Jen and Jon.

You do not have to draw people. You can use Xs or dots to stand for people in your drawing.

Work Backward

Does it ever make sense to do something backward? Think about when you get ready for bed. You take off your sneakers. Then you take off your socks. That is the opposite of what you do when you get dressed.

You can work backward to solve some problems.

You have directions to get from your house to the store. You want to tell Pete how to get from the store to your house.

You can have Pete go backward by undoing each step. What would you tell Pete to do first?

From My House to the Store

Go to the sidewalk.
Turn right.
Go 3 blocks.
Turn left.
Go 7 blocks.
Store is on the left.

If you know the end and you know the steps to get to the end, you can undo the steps to get back to the start.

Problem-Solving Strategies

Look for a Pattern

When you listen to music, you can *hear* patterns. When you walk or run, you can *feel* patterns.

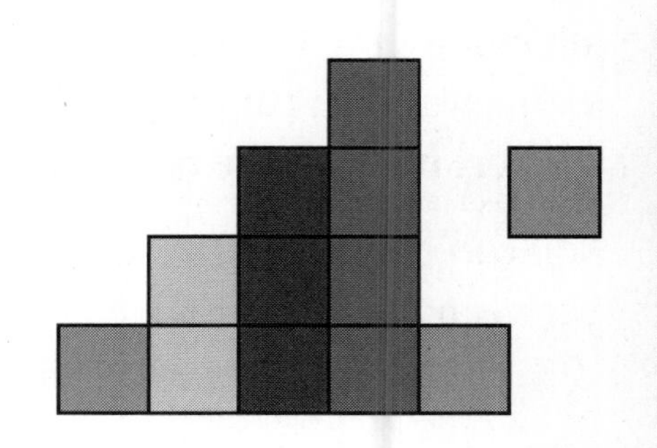

You can use numbers to describe patterns. Sometimes that is a good way to solve a math problem.

James is building steps with blocks. How many blocks does he need to make steps that are 8 blocks high?

You could find and use a pattern.

1 block high → need 1 block
2 blocks high → need 3 blocks (1 + 2)
3 blocks high → need 6 blocks (1 + 2 + 3)

Do you see how to continue the pattern to solve the problem?

Be careful with patterns. What might look like a pattern at the beginning may not keep going in the same way.

Act It Out

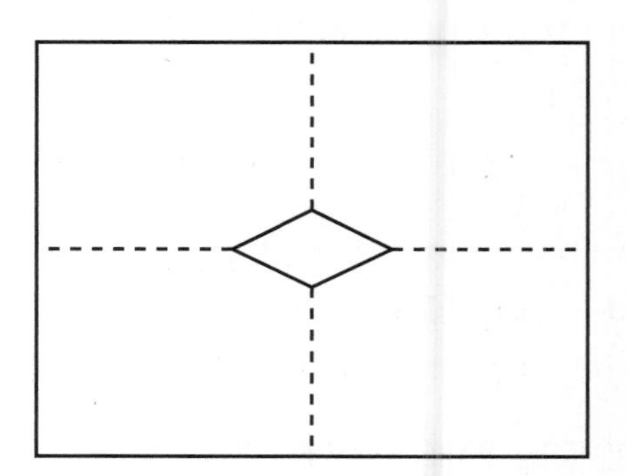

You do not have to be an actor to *act out* a math problem. You can use some objects to help you solve a problem.

Jan is going to make a design. She will fold a sheet of paper in half two times. Then she will cut along the folded corner. What will she see when she unfolds the paper?

If you want to be sure of the answer, try acting out the problem with real paper.

You do not have to use the same objects as in the problem. Suppose you are solving a problem about pianos. It is easier to use pennies than pianos.

Solve a Simpler Problem

When you first learned to ride a bike, you used training wheels. Once you got the hang of it, you did not need the extra wheels.

When the numbers in a problem seem too hard, putting in easier numbers may help you see how to solve it. Then use the same approach with the original numbers.

Mr. Smith's class collected 348 cans the first week and 765 cans the second week. Ms. Gray's class collected 627 cans the first week and 539 cans the second week. How many more cans than Mr. Smith's class did Ms. Gray's class collect?

Try solving this simpler problem first.

Class A collected 3 cans, then 7 cans. Class B collected 6 cans, then 5 cans. How many more cans than Class A did Class B collect?

A simpler problem may help you see which operations you need to use to solve the original problem.

Make a Table

With some card games, it helps to keep track of the cards. You can put the cards in rows and columns.

When you organize information in rows and columns, you make a table. You can use tables to solve math problems.

Jill feeds her cat Tank 6 ounces of food each day. After feeding Tank on Tuesday, Jill has 32 ounces of food left. When will she need to buy more cat food?

Make a table. Show how much cat food will be left each day.

Tuesday	Wednesday	Thursday	Friday
32 ounces	26 ounces	20 ounces	14 ounces

A table makes it easy to keep track of information.

Make an Organized List

Suppose you have a kit for making your own dinosaur. You can choose the head, body, and tail you want.

How many different dinosaurs can you build from the choices on the right? To find out, you can list the different dinosaurs.

Start by listing all the dinosaurs with large heads.

Large head – Green body – Spiked tail
Large head – Green body – Plain tail
Large head – Brown body – Spiked tail
Large head – Brown body – Plain tail

Then list all the dinosaurs with small heads.

Small head – Green body – Spiked tail
Small head – Green body – Plain tail
Small head – Brown body – Spiked tail
Small head – Brown body – Plain tail

Can you count all the different dinosaurs?

Build-A-Saurus

Head	Body	Tail
Large	Green	Spiked
Small	Brown	Plain

You can use letters instead of whole words. For "Large head – Green body – Spiked tail," you could just write LH – GB – ST.

Make a Graph

A graph is a kind of picture. After you gather information, you can display it in a graph. A graph shows you what your data look like.

Mrs. Green's class voted on where to go on a field trip. The graph shows the results.

Do you see how the graph makes it easy to see how the students voted?

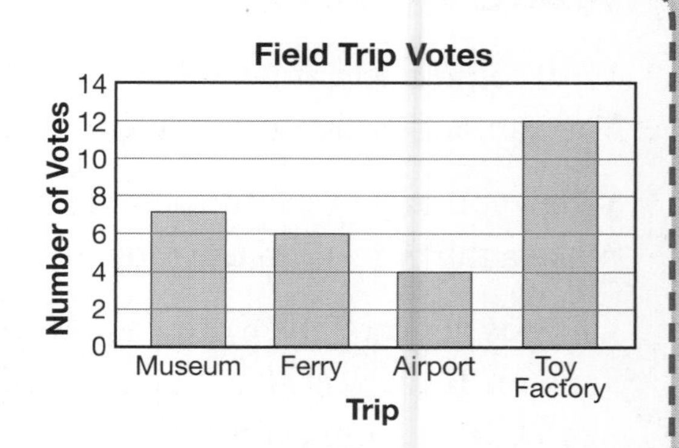

A graph can make it easier to see and compare data.

Check for Reasonable Answers

It is 3:30. Your music lesson is at 4 o'clock. You get on your bike and pedal fast. When you get to your lesson, you look at the clock. It says 8 o'clock. That cannot be right.

You know when a clock does not show the right time. You can also know when a math answer cannot be right.

Suppose you are bowling. You need to find the total score for three games. You find 85 + 93 + 78. The sum you get is 356. Is your answer reasonable?

Do you see why your answer must be less than 300?

Whenever you solve a problem, look back and think about whether your answer is reasonable. Then you can find out if something is wrong and fix it.

Decide If an Estimate or Exact Answer Is Needed

You want to buy the ball and the whistle. You have $5. There is no sales tax. Is $5 enough?

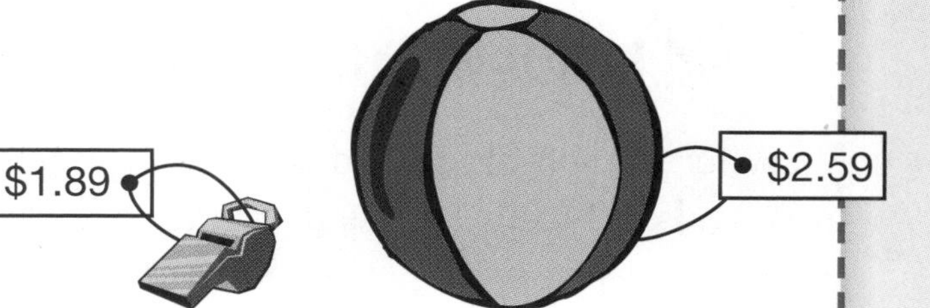

You could add $2.59 and $1.89 to find the exact total. But do you really need to?

You can estimate instead. The ball costs less than $3 and the whistle costs less than $2. Do you see why the total must be less than $5?

Sometimes you do want an exact answer. For example, when you pay for the ball and whistle with a $10 bill, you want to know *exactly* how much change you should get back.

You can use estimates to help you "check" answers. But sometimes an estimate is all you need to "find" an answer.

Decide What Information Is Unnecessary for Solving

You want to watch a TV show. You check the on-screen guide. It is full of names and numbers. You do not need all this information.

You ignore what you do not need so you can find what you are looking for. This is a very good skill to use when you solve math problems, too.

Suppose you want to know how long the show *Spell This!* lasts. Which numbers on the guide will help you?

	3:00 P.M.	3:30 P.M.	4:00 P.M.	4:30 P.M.
002 SCI	Double Discovery	Jungle Trip	On Safari	
011 KDS	Jump Start	Collect the Clue	Spell This!	Furry Friends

Solving a math problem can be like looking at a TV guide. You may not need all the information you see.

Find Hidden Information

When you play hide-and-seek, you expect your friends to hide from you. But you do not expect numbers in a math problem to hide.

Sometimes a math problem has information that is not clear at first. Maybe the information is given in a picture or in a table. Maybe the numbers you need are hidden in the words.

Matt and Dee are at the bakery. They see a muffin that costs $2. Dee says, "We can split it. I have 75 cents." "So do I," says Matt. If the two friends put their money together, can they buy the muffin?

Look carefully. Can you find where the problem tells you how much money Matt has?

You might have to read a problem more than once to find any hidden information.

Use Multiple Steps to Solve

Maybe you cannot get over both puddles in one big jump. So you take two small jumps.

To solve a math problem, you may need to do more than one step, too.

Nan has 3 large packs of stickers and 4 small packs. Each large pack has 10 stickers. Each small pack has 6 stickers. How many stickers does Nan have?

Step 1	Find how many stickers are in the large packs.	$3 \times 10 = 30$
Step 2	Find how many stickers are in the small packs.	$4 \times 6 = 24$

What will your next step be?

Your answer to a multiple-step problem depends on each step. Be sure to go over your work in each step to check your answer.

Interpret Answers

You ask your dad if you can buy some DVD movies. He says, "We will see." What does he mean?

When you are trying to figure out what something means, you are *interpreting*.

Sometimes, you need to interpret information in a math problem. You get an answer, but then you must decide what it means for you.

Each DVD costs \$3. There is no sales tax. How many DVDs can you buy with \$10?

You think: $2 \times \$3 = \$6$

$3 \times \$3 = \$9$

$4 \times \$3 = \$12$

What answer would you give?

There are different ways to interpret answers. Read the problem carefully so you know what is being asked.

Choose Strategies

Look out! Things can happen fast when you are playing dodgeball. Will you stay in the back and play it safe? Or will you use some other strategy?

You can also choose a strategy when you solve a math problem.

A fence that goes around a playground is 250 yards long. The north side is 65 yards long. The west side is 70 yards long. The south side is 50 yards long. How long is the east side?

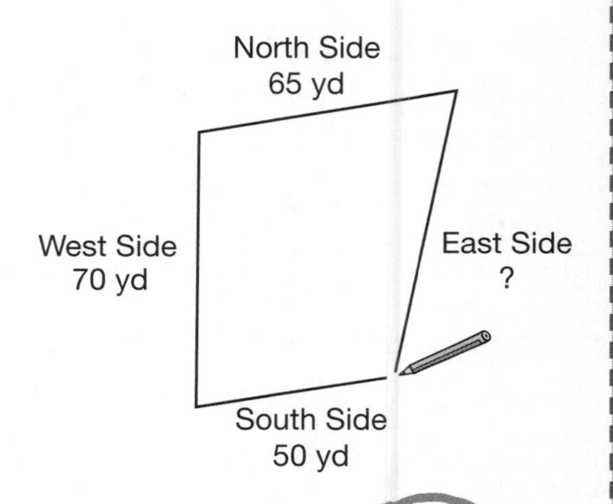

You might *draw a diagram.*

You might *write a number sentence.*
65 + 70 + 50 + east side = 250 yards

You might even use both strategies together. Or you might use some other strategy.

A great thing about solving math problems is that you get to choose the way you think works best.

Choose Operations

Dinosaurs. Skeletons. Stars and planets. Guess where you are going? That's right! You and 4 friends are going to the science museum! It costs $4 for each person. How much will it cost for all 5 of you?

You could figure out the total cost in one step. Multiply 5 × $4.

You can answer many questions with a single calculation. Be sure to think about whether to add, subtract, multiply, or divide.

When you solve a problem, always think first before computing.

Solve Two-Question Problems

When are you leaving to go to Grandma's new house? How long will it take to get there? If you want to know what time you will see Grandma, you need answers to both questions.

Some math problems ask more than one question. When they do, make sure you answer all of them.

The graph shows which pizza topping the students in Mr. Wilson's class like the most. Which topping is the favorite? How many students voted?

Two questions require two answers. Can you figure out both answers?

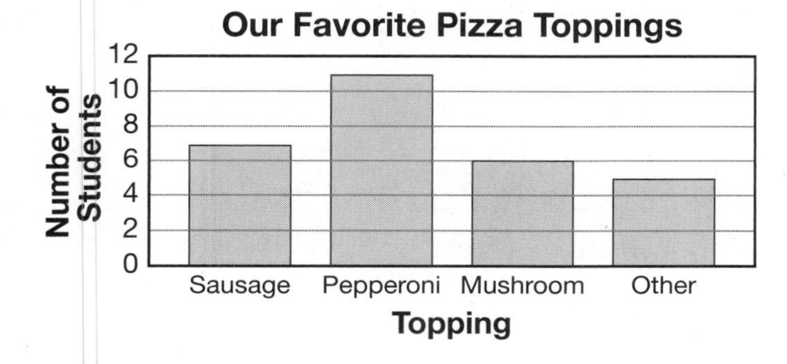

When a math problem asks more than one question, take the time to think about each question before you answer it.

Formulate Questions

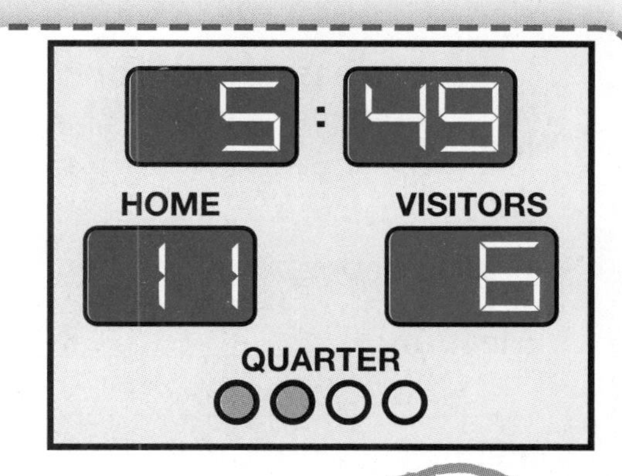

Swish! Another 3 points for the home team! The noise from the crowd gets louder and louder.

How many points is the home team ahead by? What quarter is it?

It is natural to ask questions. The better you are at asking questions, the more you can learn.

Can you think of other questions that can be answered by looking at the scoreboard?

Formulating questions about a math problem can help you understand it better.

How to Read Word Problems

Word problems usually do three things.

- They tell a story with numbers.
- They tell how the numbers go together.
- They ask a question or tell you what to find.

Read a word problem carefully to understand it. Sometimes, it helps to read it again.

Read to Understand

A word problem may have words or symbols that are new to you. Here are some things you can do.

- You can **look up** the word or symbol.
- You can use **base words.**
- You can **use context clues.**

Read the problem. Then write the meaning of each underlined word.

Mr. Brown is making ice cream. He needs 12 cups of milk. He has 2 quarts and 3 cups of milk. Use multiplication to find the number of cups of milk in 2 quarts. Add the product to 3 to find the total number of cups of milk Mr. Brown has. Does Mr. Brown have enough milk?

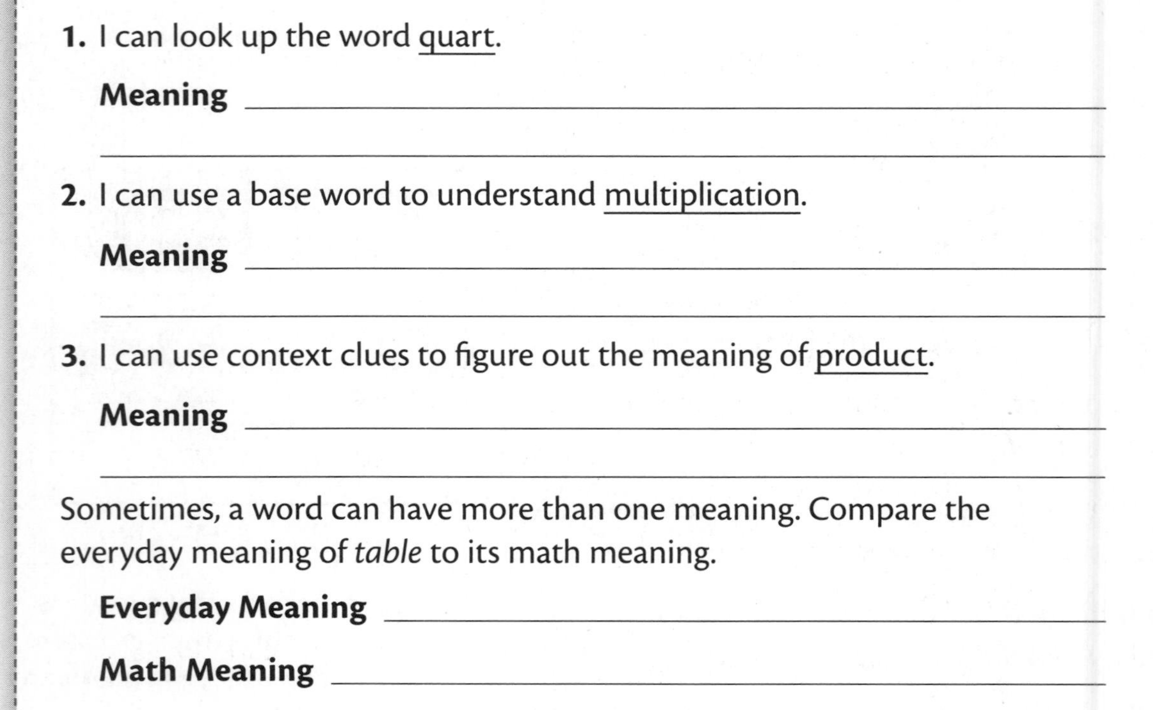

1. I can look up the word quart.

Meaning ______________________________

____

2. I can use a base word to understand multiplication.

Meaning ______________________________

____

3. I can use context clues to figure out the meaning of product.

Meaning ______________________________

____

Sometimes, a word can have more than one meaning. Compare the everyday meaning of *table* to its math meaning.

Everyday Meaning ______________________________

Math Meaning ______________________________

Look for Information

First read a word problem to get an idea of what it is about. Then read it again to find numbers and words you need to solve it.

- A problem may tell just what you need and nothing else.

  Ron is collecting cans to recycle. On Monday, Ron collected 20 cans. On Tuesday, he collected 35 cans. How many more cans did he collect on Tuesday than on Monday?

- A problem may tell everything you need and some extra details.

  On Monday, Ron collected 20 cans ~~and 14 bottles~~. On Tuesday, he collected 35 cans ~~and 28 bottles~~. How many more cans did he collect on Tuesday than on Monday?

  You do not need this information to solve the problem.

- A problem may not tell everything you need. You must look outside the problem for the rest of the information.

  Ron's garden is 20 yards long. How many feet is that?

  You need to know how many feet are in one yard.

- Sometimes, what you need is in tables, graphs, or pictures.

Read each problem below. Decide what will help you solve the problem. Then write your answer.

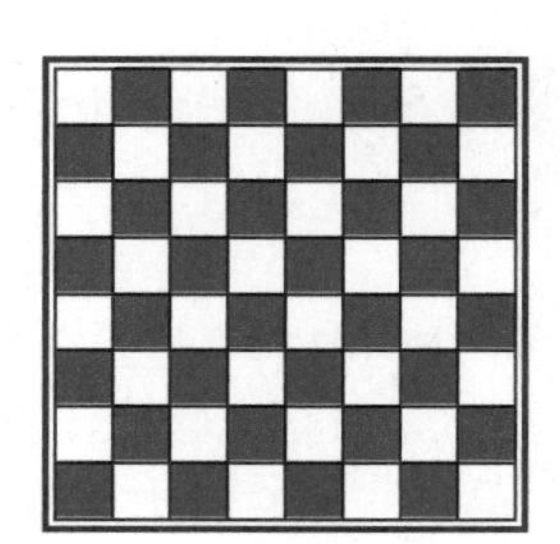

1. To start a game of checkers, Em uses the first 3 rows of the checkerboard nearest her. She puts 1 checker on each blue square in those 3 rows. How many checkers does Em use?

Information I can get only from the picture: ____________________

____

2. Your classmates made a table to show the pets they have. How many more dogs than cats do your classmates have?

Our Pets

Pet	Dog	Cat	Fish	Bird
Number	15	12	4	8

Information I can get only from the table: ____________________

____

Mark the Text

You can mark words and numbers in a problem. This will help you decide how to solve the problem.

- You can circle numbers, which can be written as numerals or words.
- You can cross out information you do not need.
- You can underline the question.
- You can mark something you need to look up.

Your dog eats 5 pounds of food each week. ~~His food costs $6 for 10 pounds~~. How many pounds of food does your dog eat in one year?

Mark the text and tell why each mark is important.

Bella has a cat and a dog. Her cat's food costs $2 for one week. Her dog's food costs $5 for one week. How much does it cost to feed Bella's dog for a year?

I underlined ______________________________

because ______________________________

I crossed out ______________________________

because ______________________________

I circled ______________________________

because ______________________________

Decide What to Do

After you mark the text, think about how you will solve the problem.

- Sometimes, you can solve a word problem by computing.

  On Monday, Kara counted 20 cars parked on her block. On Tuesday, she counted 18 cars. Today is Wednesday and Kara has counted 13 cars. How many cars has Kara counted so far this week?

Explain What operation or operations will you use? Explain how you decided.

- Sometimes, the answer you get when you compute is not the answer to the question.

  On Monday, Kara counted 20 cars parked on her block. On Tuesday, she counted 18 cars. Today is Wednesday and Kara has counted 13 cars. Has Kara counted more than 50 cars so far this week?

Determine Why is the sum not the answer to this question?

- Sometimes, you do not need to compute at all. You can find the answer a different way.

  On every day that Kara sees more than 15 cars, she calls her friend Ella. Name the days that Kara calls Ella.

Sunday	Monday	Tuesday	Wednesday	Thursday	Friday	Saturday
0	20	18	13	12	16	30

Apply There is no question here. What do you need to do?

Problem Solving Using Place Value, Addition, and Subtraction

Unit Theme: Trips

Have you been in a boat? Do you like to visit a theme park? Maybe you dream of riding in a hot air balloon. Trips to new places can be full of fun. In this unit, you will see how math is used in many places.

Math to Know

In this unit, you will use these math skills:

- Use place value and compare numbers
- Round numbers
- Add and subtract whole numbers

Problem-Solving Strategies

- Make a Table
- Write a Number Sentence
- Draw a Picture
- Work Backward

Link to the Theme

Finish the story. Include some of the facts from the table at the right.

Mason and his dad are at the Wild Waves Water Park. They stop to read a sign. It lists facts about three water slides.

____

____

____

____

____

____

____

Wild Waves Water Slides

Name	Greatest Height	Length
Splash Path	98 feet	419 feet
Storm Hill	112 feet	507 feet
Tide Ride	87 feet	434 feet

Use Math Language

Review Vocabulary

Here are some math words for this unit. Knowing what these words mean will help you understand the problems.

addend	difference	number sentence	round
compare	estimate	place value	sum

Vocabulary Activity **Words in Context**

Context can help you find out what a word means. Use words from the list. Complete each sentence.

1. The ____________________ of the 3 in the number 35 is tens.
2. A number that is being added is called an ____________________ .
3. When you add two numbers, you find their ____________________ .
4. To make an ____________________ , look for a number that is close to an exact number.

Graphic Organizer **Word Map**

Complete the graphic organizer.

- Draw a picture or diagram to show what the term means.
- Write a number sentence that shows a difference.
- Write a number sentence that does not show a difference.

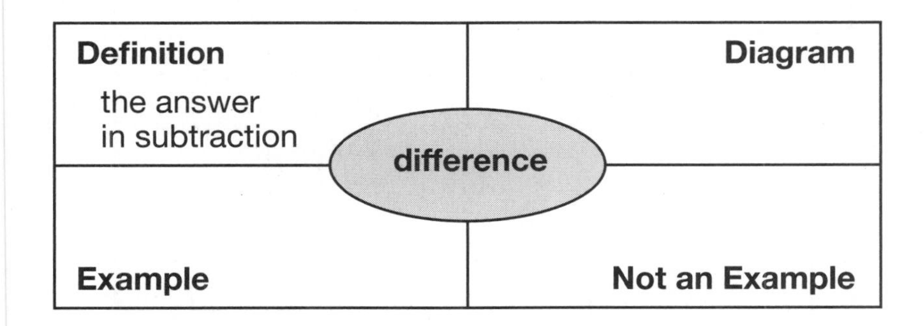

Strategy Focus

Make a Table

MATH FOCUS: Place Value and Numeration

Learn About It

Read the Problem

Ms. Lee's class is learning about dinosaurs. Triceratops weighed about 13,450 pounds. Stegosaurus weighed about 6,060 pounds. T. Rex weighed about 13,990 pounds. Ms. Lee asks the class to make a poster about the dinosaurs. They need to show the dinosaurs in order from least to greatest weight. What should the order of dinosaurs be?

Reread Look for important facts.

- What is this problem about?

____

- What does the problem ask you to do?

____

- What will you compare to solve the problem?

____

Mark the Text

Search for Information

Read the problem again. Circle words and numbers that will help you solve the problem.

Record Write the weight for each dinosaur.

Triceratops weighed about ________________ pounds.

Stegosaurus weighed about ________________ pounds.

T. Rex weighed about ________________ pounds.

Use these facts to help you solve the problem.

Decide What to Do

You know the weights of three kinds of dinosaurs. You need to put the weights in order.

Ask How can I put the weights in order from least to greatest?

- I can use the strategy *Make a Table* to compare the weights.

Use Your Ideas

Step 1 Make a table. Write each digit of the weights in the column that matches its place value.

	Place Value				
	Ten Thousands	Thousands	Hundreds	Tens	Ones
Triceratops	1	3	4	5	0
Stegosaurus					
T. Rex					

The digit in the Thousands column is the same for Triceratops and T. Rex.

Step 2 Compare the weights. Start at the left. The weight for Stegosaurus has no digit in the Ten Thousands column. So it has the least weight. The least weight is ________________ pounds.

Step 3 Compare the weights of the other two dinosaurs. Find the column where the digits are different.

The greatest weight is ________________ pounds.

The order of dinosaur weights in pounds from least to greatest is __________, __________, and __________.

The order of dinosaurs from least to greatest weight is __________, __________, and __________.

Review Your Work

Look back. Check that you used the right numbers.

Describe How does the table help you compare the numbers?

____

____

Try It

Solve the problem.

① A dinosaur park needs a new sign. It will show the weights of two dinosaurs. The weights will be rounded to the nearest hundred. The old sign is on the right. What weights will be given on the new sign?

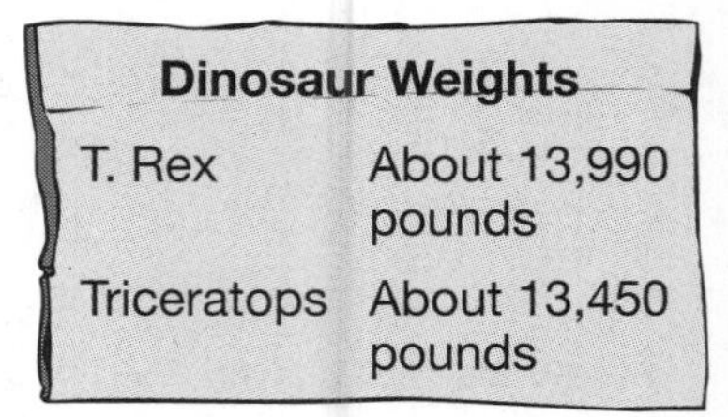

Dinosaur Weights	
T. Rex	About 13,990 pounds
Triceratops	About 13,450 pounds

Read the Problem and Search for Information

Circle facts that will help you solve the problem.

Decide What to Do and Use Your Ideas

You can use the strategy *Make a Table.*

Step 1 Make a table. Write the numbers in the table.

	Place Value				
	Ten Thousands	**Thousands**	**Hundreds**	**Tens**	**Ones**
T. Rex	1	3	9	9	0
Triceratops					

Ask Yourself

What place value do I round to?

Step 2 To round each number, look at the digit to the right of the ______________ place.

Step 3 13,990 rounded to the nearest hundred is ________.
13,450 rounded to the nearest hundred is ________.

The new sign will show that T. Rex weighs about ________ pounds and Triceratops weighs about ________ pounds.

Review Your Work

Check that you rounded to the correct place value.

Recognize What word tells you that rounded numbers are being used?

____

____

Apply Your Skills

Solve the problems.

2. There are two T. Rex fossils at a museum. One is called Betty and the other is called Barney. Betty is 570 inches long. Barney is 539 inches long. Which fossil is longer? Which place value did you use to decide?

	Place Value		
	Hundreds		
Betty			
Barney			

Hint Fill in the missing place-value names in the table.

Answer ______________________________

Explain Why do you need to start at the left when you compare the digits instead of starting at the right?

____

____

Ask Yourself

Start at the left. What is the place value of the first digits that are different?

3. In June, 13,832 people came to a dinosaur park. In July, 14,186 people came to the park. Round the number of people to the nearest thousand for each month. Did about the same number of people visit the park each month? How do you know?

	Place Value				
		Thousands		Tens	
June					
July					

Hint Fill in the missing place-value names. Then write the numbers in the table.

The rounded number for June is ________________ .

The rounded number for July is ________________ .

Answer ______________________________

Identify If you round to the nearest thousand, what is the least number of people that would round to 15,000?

____

____

Ask Yourself

What place should I look at to help me round to the nearest thousand?

Hint Do not skip places when you compare.

Ask Yourself

Which digits should I compare first?

4 There are 2 hiking paths at a dinosaur park. The Red Path leads to some three-toed tracks. This path is 12,198 feet long. The Green Path leads to some round tracks. That path is 12,402 feet long. Which path is shorter?

	Place Value				
Red Path					
Green Path					

Answer ______________________________

Apply Explain how you used place value to decide which path is shorter.

____

____

Hint You need to compare the number of votes for three dinosaurs.

Ask Yourself

Which place-value names should I write in the table?

5 The dinosaur park had a contest. People voted for their favorite dinosaur. T. Rex got 1,679 votes. Stegosaurus got 1,684 votes. Spinosaurus got 1,681 votes. Which dinosaur got the most votes?

Answer ______________________________

Determine Suppose Triceratops was also in the park's contest. It got 1,680 votes. How would you change the table to show the number of votes for Triceratops?

____

____

On Your Own

Solve the problems. Show your work.

6. The Dinosaur Gift Shop has 3 shelves of toys. Last year the shop sold 4,835 toys. This year, the shop sold 4,829 toys. Did the shop sell more toys last year or this year?

Answer ______________________________

Analyze What number in the problem is not needed to solve it?

____

____

7. A dinosaur park has a snack bar. It sold 316,586 bottles of water last year. It sold 314,958 bottles this year. To decide how many bottles to order next year, the greater number is rounded to the nearest ten thousand. How many bottles should be ordered next year?

Answer ______________________________

Evaluate Why would rounding to the nearest hundred thousand not make sense?

____

____

Create

Look back at Problem 5. Change the number of votes for each dinosaur so that your favorite is the winner. Write a problem about your contest. Solve your problem.

Strategy Focus

Write a Number Sentence

MATH FOCUS: Addition

Learn About It

Read the Problem

Kari and her sister Ann went on a 5-mile boat trip. They rowed for 45 minutes. Then they stopped to have a snack. After their snack, Kari and Ann rowed another 20 minutes before stopping for lunch. How many minutes in all did they row before lunch?

Reread Read the problem carefully.

- What question is the problem asking?

____

- Which fact will *not* help you answer the question?

____

Mark the Text

Search for Information

Think about the question in the problem. Circle words and numbers that will help you answer that question.

Record What facts do you know?

They rowed for ________ minutes until they stopped for a snack.

Then they rowed for another ________ minutes before stopping for lunch.

Think about how you can use these numbers to solve the problem.

Decide What to Do

The question asks how many minutes Kari and Ann rowed in all. You know the two different numbers of minutes they rowed.

Ask How can I find how many minutes Kari and Ann rowed before lunch?

- I can use the strategy *Write a Number Sentence.*
- I can write an addition sentence.

You are asked to find a total.

Use Your Ideas

Step 1 To write a number sentence, decide whether each number is an addend or a sum.

The time from the start until the snack is an addend.

The time from the snack until lunch is an ___________.

The time the sisters rowed in all before lunch is the sum.

Step 2 Write an addition sentence in words to tell what you know. Next, write a number sentence. Then find the missing number.

Minutes rowed until snack	+	Minutes rowed after snack	=	Minutes in all before lunch
45	+	20	=	________

So Kari and Ann rowed ________ minutes before lunch.

Review Your Work

Look back at the problem. Does your answer make sense?

Explain Why can you use an addition sentence to solve this problem?

____

____

Try It

Solve the problem.

① Yuko and his father went on a road trip. The trip was 250 miles long. It took 2 days. On the first day, they went 200 miles. How far did they go on the second day?

Read the Problem and Search for Information

Find the question you have to answer. Circle the numbers you need to use.

Decide What to Do and Use Your Ideas

You can write a number sentence to show what happens.

Step 1 Write what you know.

They went ________ miles on the first day.

They went ________ miles in all.

Step 2 Write a number sentence. Use a question mark (?) for what you need to find.

Miles on the first day	+	Miles on the second day	=	Total miles
________	+	________	=	________

Step 3 Find the missing number.

200 + ________ = 250

Ask Yourself

How can I use mental math to find the missing number?

They traveled ________ miles on the second day.

Review Your Work

Put your answer in the number sentence. Then check to make sure that the sentence is correct.

Recognize Do you need to find a missing sum or a missing addend to solve this problem? Explain how you know.

____

____

Apply Your Skills

Solve the problems.

② Sara and Drew keep track of how many miles they ride their bikes. So far, Drew has gone 320 miles. Sara has gone 70 miles more than Drew. How many miles has Sara gone?

Hint Write a number sentence in words. Think about what you know and what you must find.

Number of Drew's miles	+	Number of miles more Sara has gone	=	Number of Sara's miles
320	+	________	=	________

Ask Yourself

What number can I add to 320 to find the answer?

Answer ______________________________

Describe Explain how to estimate to check your answer.

____

____

③ Ms. Wang sells rides on a hot air balloon. On Monday and Tuesday, she sold a total of 17 rides. On Monday she sold 9 rides. How many rides did she sell on Tuesday?

Ask Yourself

Which number is the sum?

Number of rides on Monday	+	Number of rides on Tuesday	=	Total number of rides
________	+	________	=	________

Hint You need to find a missing addend.

Answer ______________________________

Contrast What is another number sentence you could use to solve this problem?

____

____

4 Bill and Tom like to hike. Tom hiked 40 kilometers more than Bill. Tom hiked 100 kilometers. What was Bill's hiking distance?

Ask Yourself

How do the words *more than* help me?

Hint Think of the number that you need to add to 40 to get a sum of 100.

__________ + Number of kilometers more that Tom hiked = Tom's hiking distance

______ + ______ = ______

Answer ____________________

Conclude How did you decide how to write your number sentence?

____

____

5 The Jenks family rode in a train. They rode 30 miles and stopped at Old West Town. Then they rode another 15 miles on the train to a ranch. On the train ride back, they traveled on the same tracks through Old West Town. How many miles did the family ride on the train?

Ask Yourself

How many miles did they travel on the ride back?

Hint You can write an addition number sentence with four addends.

______ + ______ + ______ + ______ = ______

Answer ____________________

Evaluate What is another strategy that you could use to help you solve this problem?

____

____

On Your Own

Solve the problems. Show your work.

6. Jean takes horseback riding lessons. Her goal is to ride 150 miles this year. So far, Jean has ridden 90 miles. How many more miles will she need to ride to reach her goal?

Answer ______________________________

Compare How is this problem like Problem 1?

____

____

7. Lela skates in a race. The first part of the race is 420 feet. The second part of the race is 50 feet longer than the first part. What is the total distance Lela skates in the race?

Answer ______________________________

Justify How do you know that your answer is correct?

____

____

Create

Look back at Problem 4. Change the names of the friends and change the distances each friend hiked. Write and solve a new problem. Then solve your problem.

Lesson 3

Strategy Focus

Draw a Picture

MATH FOCUS: Subtraction

Learn About It

Read the Problem

Cam's uncle hid a gift for Cam to find. He told Cam to start at a big rock and walk 87 yards north. Then he said to turn east. His uncle did not remember how many yards east to walk. He said to walk 200 yards altogether. How many yards east should Cam walk?

Reread Think about what is happening in the problem.

- What is the problem about?

____

- In what two directions will Cam walk?

____

- What do you have to figure out?

____

Search for Information

Mark the details that will help you solve the problem.

Record Write the order in which Cam should walk and the number of yards he should walk. Use a question mark for anything you do not know.

- Cam should start at the big ________________.
- Cam should walk ________ yards north.
- Cam should turn east and walk ________ yards.
- Cam should walk ________ yards altogether.

Think of a strategy that will help you see what is happening in the problem.

Decide What to Do

You know that Cam should walk north and east. You know how many yards he should walk north. You also know the total number of yards he should walk.

Ask How can I find how far east Cam should walk?

- I can use the strategy *Draw a Picture* to show Cam's path.
- I can use the picture to decide what operation to use.

Use Your Ideas

Step 1 Draw a picture to show Cam's path. Put in all the information you know. Use a question mark for anything you do not know yet.

Number of yards: ________

Direction to walk: ________

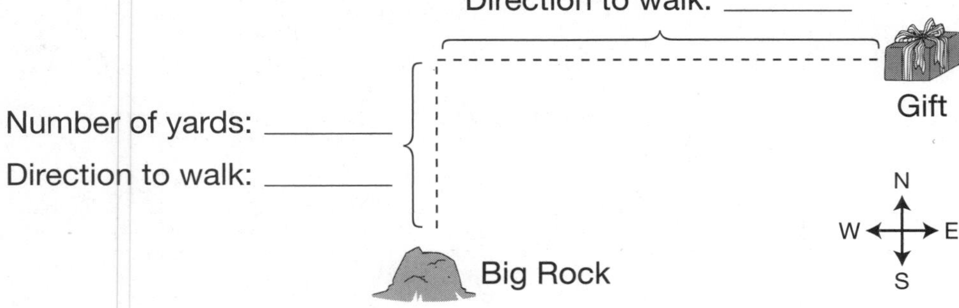

Step 2 Write a number sentence. Show how the two parts of the path make the whole path.

Find the difference between the whole and the unknown part.

200 − ________ = ________

So Cam should walk ________ yards east.

You know the whole or total. You can subtract to find an unknown part.

Review Your Work

Use addition to check your answer. The sum should be 200.

Compare How are the picture you drew and the number sentence alike?

____

____

Try It

Solve the problem.

① A park has a path that is 150 yards long. The first part is from the park gate to the swings. The second part is 50 yards to the snack bar. The third part is 38 yards to the exit. How many yards is it from the gate to the swings?

Read the Problem and Search for Information

Mark details about directions and distances.

Decide What to Do and Use Your Ideas

You can use the strategy *Draw a Picture* to show the path.

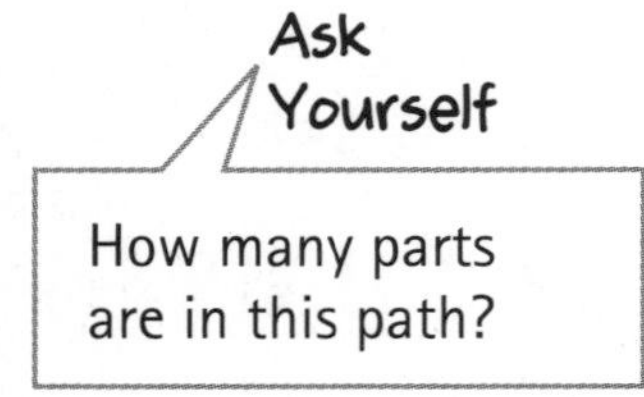

Step 1 Draw the path. Write the length of each part you know.

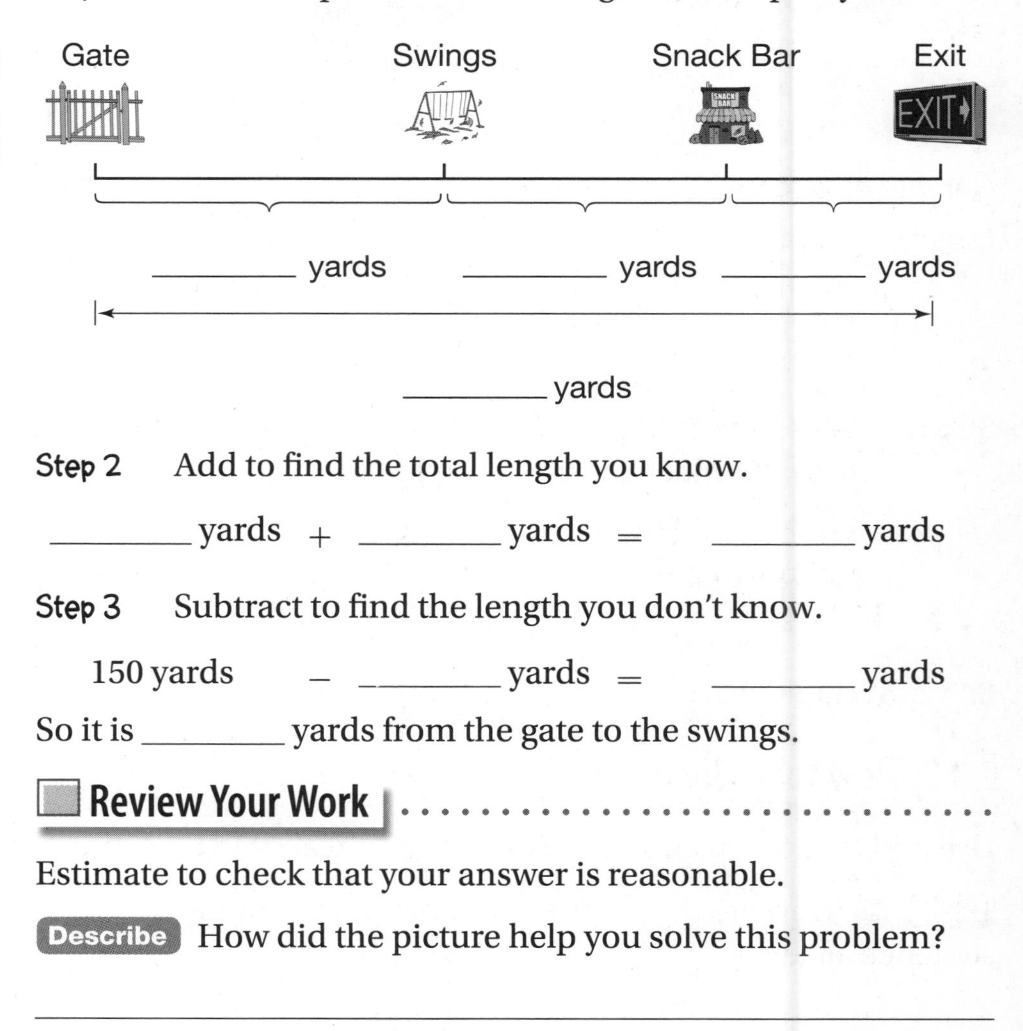

Step 2 Add to find the total length you know.

______ yards + ______ yards = ______ yards

Step 3 Subtract to find the length you don't know.

150 yards − ______ yards = ______ yards

So it is ______ yards from the gate to the swings.

Review Your Work

Estimate to check that your answer is reasonable.

Describe How did the picture help you solve this problem?

____

____

Apply Your Skills

Solve the problems.

② Miko walks around two sides of her yard. The picture shows how she walked. Her path is 323 feet long. Her brother takes another path across the yard. His path is 247 feet long. How much farther does Miko walk than her brother?

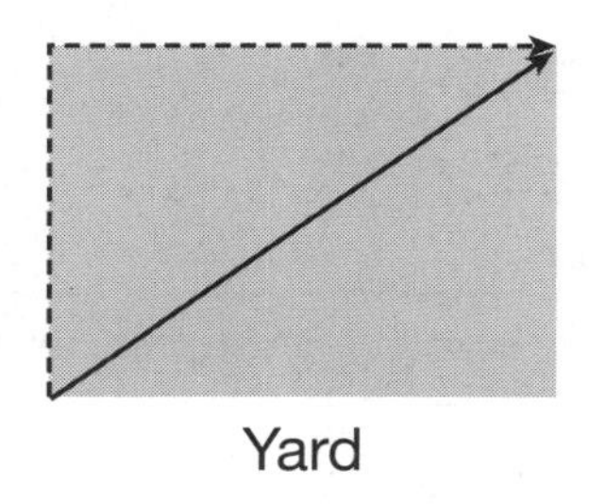

Hint Use the picture to compare the paths.

Ask Yourself Which operation should I use?

Miko's path is ________ feet long.

Her brother's path is ________ feet long.

Answer ______________________________

Explain Make an estimate to help you check that your answer is reasonable. Explain how you made it.

____

____

③ Doug told Mario about a new store he found in town. Doug said, "I started at my house. Then I went 526 yards north to get to the new store." Mario's house is 342 yards north of Doug's house. How far is the store from Mario's house?

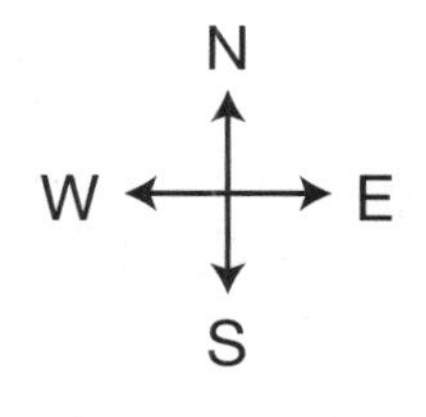

Ask Yourself Can I draw a picture for this problem?

The distance from Doug's house to the store is ________ yards.

The distance from Doug's house to Mario's house is ________ yards.

Hint You can write a number sentence to help you.

Answer ______________________________

Conclude What details in the problem tell you that Doug passed Mario's house on his way to the store?

____

____

Ask Yourself

How does the picture help me decide which operations to use?

4 Karen walks to a picnic along the trail shown in the picture. She follows the trail for 999 meters. First, she walks 208 meters to a big rock. Then she walks to a sign. At the sign, she walks another 292 meters to the picnic. How many meters is it from the big rock to the sign?

Karen follows the trail for ________ meters.

Start Rock Sign Picnic

Hint Find the unknown part. ▶

________ meters ________ meters ________ meters

Answer ________________________________

Interpret How does the total distance drawn at the top of the picture help you solve the problem?

____

____

5 Sam walked a total of 500 feet to a pool. He walked 236 feet west. Then he walked 90 feet south. This was the wrong direction. So he walked north. How many feet north did he walk to reach the pool?

N
W E
S

Hint Sam has to walk back on the same path he walked south. ▶

Sam walked a total of ________ feet.

He walked ________ west.

Then he walked ________ feet south.

Ask Yourself

How can I show the different directions in my picture?

Answer ________________________________

Revise Use the information in the problem. Find a shorter way to the pool. What is the length of the shorter way?

____

____

On Your Own

Solve the problems. Show your work.

6. Willa found a treasure map. It said, "Start where you found this map. Walk 235 steps east. Then walk 419 steps west." Willa figured out that she could just go west to get to the treasure faster. How many steps west should she go?

Answer ______________________________

Determine Why can Willa find the treasure by just going west?

____

____

7. Dom and Jon are at the mall. They want to go to the toy store on the third floor. First, they go up 12 stairs to get from the street to the first floor of the mall. They go up 24 stairs from the first floor to get to the second floor. Then they go up more stairs from the second floor to get to the third floor. Dom and Jon climb 60 stairs in all. How many stairs do they climb from the second floor to the third floor?

Answer ______________________________

Consider What is another question this problem could ask?

____

____

Create

Look back at Problem 6. Make up a problem about a treasure map. Have the person following the map go in at least 3 different directions. Give the total distance the person needs to walk. Give the distances for all but one of the directions. Ask what the missing distance is. Solve your problem.

Lesson 4

Strategy Focus Work Backward

MATH FOCUS: Addition and Subtraction

Learn About It

Read the Problem

A spaceship left Planet X with a certain amount of fuel. It used 185 gallons of fuel to get to Planet Z. At Planet Z, the spaceship added 60 gallons of fuel. Then it had 275 gallons of fuel in its tank. How much fuel did the spaceship have at the start?

Reread Think about the parts of the trip. Look for the question you need to answer.

- What is the problem about?

____

- What do I need to find?

____

Mark the Text

Search for Information

Read the problem again. Circle the numbers that you will need to solve the problem.

Record Write the information you will use.

The spaceship had ________ gallons of fuel at the end of the trip.

The spaceship added ________ gallons of fuel at Planet Z.

The spaceship used ________ gallons of fuel to get to Planet Z.

You know the end but you do not know the start. This information can help you choose a problem-solving strategy.

Decide What to Do

You know how much fuel the spaceship had at the end of the trip. You know how much fuel the spaceship used. You know how much fuel the spaceship added.

Ask How can I find the amount of fuel at the start?

- I can use the strategy *Work Backward.* I can begin with the end to find how much fuel the spaceship had at the start.

Use Your Ideas

Step 1 Draw a diagram to show what happens.

Subtract 185 gallons. → Add 60 gallons. →

Start	?		?		275	End

Step 2 Draw a diagram to show working backward. Record your answers in the diagram below.

Start					275	End

← Add 185 gallons. ← Subtract 60 gallons.

First, subtract the fuel the spaceship got on Planet Z.

Then add the fuel the spaceship used to get from Planet X to Planet Z to your answer from above.

So the spaceship started with ________ gallons.

To work backward, undo what happened in the problem. To undo an operation, use the opposite operation.

Review Your Work

Start with your answer. Use it in the first diagram. Do you have 275 gallons at the end?

Recognize Why is the strategy *Work Backward* a good one for solving this problem?

____

____

Try It

Solve the problem.

① Sophie visits Planet Zorg. She finds 3 new kinds of rocks. She calls them muck rocks, slab rocks, and plop rocks. Sophie finds some muck rocks on the first day. The next day, she finds 10 slab rocks and 32 plop rocks. She finds 66 rocks in all. How many muck rocks does she find?

Read the Problem and Search for Information

Find the question. Circle the numbers you need to use.

Decide What to Do and Use Your Ideas

You can use the strategy *Work Backward.* You can start with what you know about the end. You can work back to the start.

Ask Yourself

Which operation or operations will I use to solve this problem?

Step 1 Draw a diagram to show what happens.

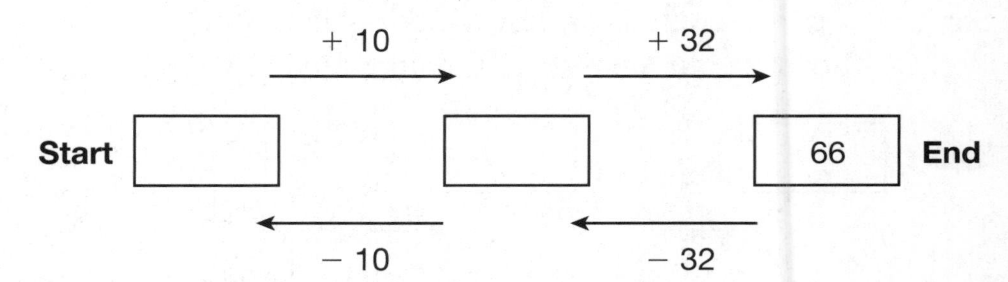

Step 2 Use opposite operations to work backward.

Write your answers in the diagram above.

First, subtract 32 from 66. $66 - 32 =$ ________

Then subtract 10 from that answer.

________ $- 10 =$ ________

So Sophie finds ________ muck rocks.

Review Your Work

Make sure that you answered the question that was asked.

Tell How you can check that your answer is correct?

____

____

Apply Your Skills

Solve the problems.

② Ko explores the Planet Zib. He looks for samples to test. First, he finds some seeds. Then he finds 6 kilograms of sand and 13 kilograms of stones. The three samples have a total mass of 23 kilograms. What is the mass of the seeds?

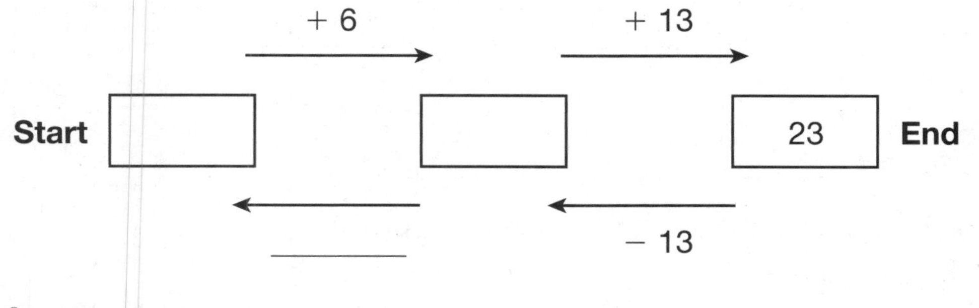

Answer ____________________

Describe You can subtract twice to solve this problem. What is another way to solve it?

____

____

Ask Yourself

Which operation or operations will I use to work backward?

Hint The mass of the seeds is less than the total mass of the three samples.

③ A spaceship stops at two planets. At Planet Vend, 75 people get off. At Planet Tron, 64 people get on. There are now 176 people on the spaceship. How many people were on the spaceship before the two stops?

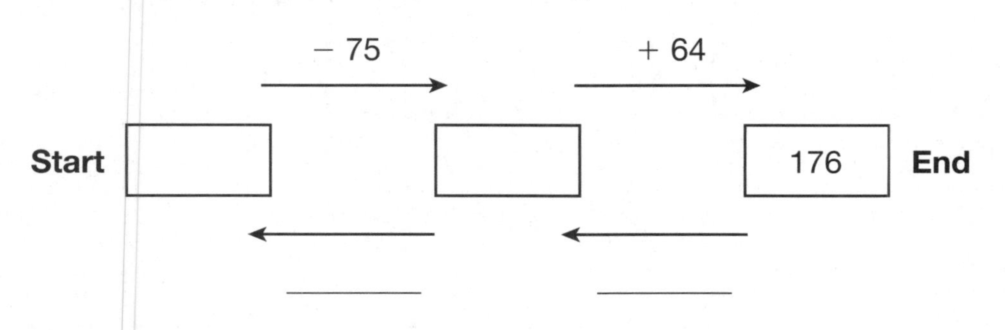

Answer ____________________

Explain How would you solve the problem if no one got on or off the spaceship at Planet Tron?

____

____

Hint The spaceship started with more people than it has now.

Ask Yourself

How many operations will I use to get back to the start?

Hint There are 60 minutes in 1 hour.

4. Bo lives on a space station. He exercises 2 hours every day to stay fit. First, Bo rides an exercise bike. Next, he runs on a treadmill for 35 minutes. Then he uses weights for 55 minutes. How long does Bo ride the bike?

Ask Yourself

What should all the minutes add up to?

+ 35 →　　+ 55 →

Start [ ] [ ] [ ] **End**

← ______　← ______

Answer ________________________

Determine How does the diagram help you solve the problem?

____

____

Ask Yourself

What number do I use for the end?

5. A spaceship leaves Planet Zab with a certain amount of water. The crew uses 480 gallons of water on the way to Bot. They get 250 gallons of water on Bot. Then they use 550 gallons on the way back to Zab. When they get back to Zab, they have no water left. How many gallons of water did the spaceship have when it left Planet Zab?

Hint Use the diagram to keep track of how much water is on the spaceship.

______ →　______ →　______ →

Start [ ] [ ] [ ] [ ] **End**

← ______　← ______　← ______

Answer ________________________

Identify Tell another way you can solve this problem.

____

____

On Your Own

Solve the problems. Show your work.

6. Zork works in a space supply room. Only space suits, helmets, and air tanks are in the room. Zork knows that there are 128 items in the room. There are 62 helmets and 37 air tanks. How many space suits are in the supply room?

Answer ______________________________

Formulate What other question can you ask using the information in this problem?

____

____

7. On Planet Nod, Dr. Stone picked up 18 sand samples. On Planet Neet, she picked up 9 sand samples. On Planet Mot, she found 27 sand samples. She now has 64 sand samples in all. How many samples did she have before she went to Planet Nod?

Answer ______________________________

Discuss How can you work backward to solve this problem?

____

____

Create

Look back at Problem 3. Write a problem about stopping at other planets. How many people get on and off at each planet? Solve your problem.

UNIT 1 Review What You Learned

In this unit, you worked with four problem-solving strategies. You can often use more than one strategy to solve a problem. So if a strategy does not seem to be working, try a different one.

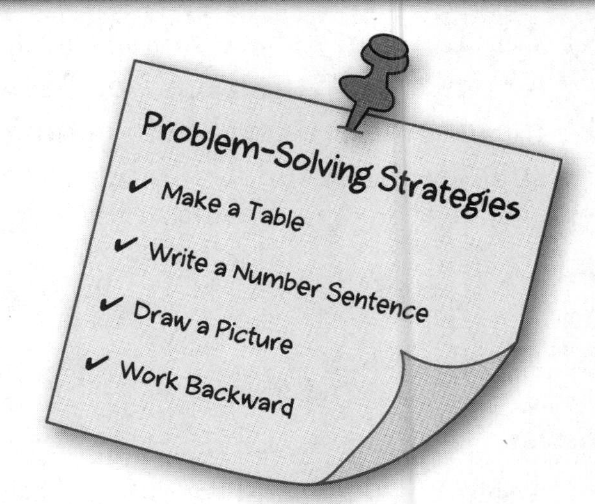

Solve each problem. Show your work. Record the strategy you use.

1. Becky and Brianna want to count their steps. They each use a tool that counts their steps. Becky took 10,859 steps on Saturday. Brianna took 11,012 steps. Who took more steps?

Answer ____________________

Strategy ____________________

2. Mr. Garcia drove 132 miles on Monday. On Tuesday, he drove 17 more miles than on Monday. How many miles did he drive on Tuesday?

Answer ____________________

Strategy ____________________

3. Ms. Lee picked 83 tomatoes from her garden. She gave away 45 of them. She also used 12 of them to make sauce. How many tomatoes does she have left?

Answer ______________________________

Strategy ______________________________

4. Greentown had three events for Earth Day. The mayor planted a tree at City Hall in front of 20,385 people. At the town cleanup, 20,419 people picked up trash. Later, 20,320 people watched a concert. List these events in order from greatest to least number of people.

Answer ______________________________

Strategy ______________________________

5. Pedro collects baseball cards. On Thursday, he bought 38 cards. A friend gave him 10 more. Now he has 423 baseball cards. How many baseball cards did Pedro have on Wednesday?

Answer ______________________________

Strategy ______________________________

Explain how you solved the problem.

Solve each problem. Show your work. Record the strategy you use.

6. Kyle takes his 3 sisters for wagon rides around their block. They will go 660 yards. Two sisters fit in the wagon at a time. The third sister walks beside Kyle. Abby walks 235 yards. Then Cara walks. Mira walks the last 189 yards. How far does Cara walk?

Answer ______________________________

Strategy ______________________________

7. Ms. Walters drives a bookmobile. At her first stop on Monday, 34 books were returned. Then 78 books were checked out. That left exactly 500 books in the truck. How many books did Ms. Walters start with on Monday?

Answer ______________________________

Strategy ______________________________

8. Sage, Maya, Jeff, and Abu have saved empty cans to recycle. Sage has saved 46 cans. Maya has saved 27. Together Jeff and Abu have saved the same number as Sage and Maya. Abu has saved 38 cans. How many has Jeff saved?

Answer ______________________________

Strategy ______________________________

Explain the steps you took to find your answer.

9. Nadia walked a total of 225 meters to the store. She walked south first. Then she walked 40 meters east. Finally, she walked 53 meters south. How many meters south did Nadia walk?

Answer ____________________

Strategy ____________________

10. Pete's dog Lucy found some tennis balls on Wednesday. She found 38 more on Thursday. Lucy lost 19 balls on Friday. They still have 36 tennis balls. How many balls did Lucy find on Wednesday?

Answer ____________________

Strategy ____________________

Write About It

Look back at Problem 6. Explain why you used the strategy you did to solve the problem.

____

____

Work Together: Choose Your Prizes

Each person on your team earns tickets playing a game. You and the team must decide which prizes to choose using your tickets.

Prize	Tickets
Eraser	5
Pencil	10
Notebook	25
Balloon	50
Sunglasses	100
Travel cup	250

Plan

1. Each team member writes a 3-digit number. Put the numbers in one pile and mix them up. Each player draws one of the numbers. The number you draw is the number of tickets you have.
2. Choose prizes. Make sure you have enough tickets. You can share tickets with other team members to buy prizes worth a lot of tickets.
3. Discuss your choices as a team. Show how you knew you had enough tickets for each prize.

Evaluate Team members should check each other's work.

Organize Find an organized way to show everyone's choices.

Present Share you team's results with the class.

UNIT 2 Problem Solving Using Multiplication and Division

Unit Theme: Nature

Nature is all around you. Look closely. You can see all kinds of things! You might see birds looking for food or ants marching along. There might be plants growing or bees flying by. In this unit, you will see how math is in nature, too.

Math to Know

In this unit, you will use these math skills:

- Multiply and divide using basic facts
- Multiply by 1-digit numbers
- Divide by 1-digit divisors

Problem-Solving Strategies

- Look for a Pattern
- Write a Number Sentence
- Act It Out
- Solve a Simpler Problem

Link to the Theme

Finish the story. Include numbers and some of the words at the right.

Cal and his family spend the day at the beach. They see many shells. Cal and his sister count and share the seashells.

Words to Use

equal	how many
divide	multiply
groups	total

____

____

____

____

____

____

____

Use Math Language

Review Vocabulary

Here are some math words for this unit. Knowing what these words mean will help you understand the problems.

array	equal groups	multiply	product
divide	factor	pattern	rule

Vocabulary Activity **Multiple-Meaning Words**

Some words have more than one meaning. Use two words from the list above to complete the following sentences.

1. The store gave away free samples of the new cleaning __________________ .
2. The answer to a multiplication problem is called a __________________ .
3. A __________________ is a way to describe a pattern.
4. A __________________ tells you what you can and cannot do in a game.

Graphic Organizer **Word Circle**

Complete the graphic organizer.

- Cross out the word that does not belong.
- Replace it with a word from the vocabulary list that does belong.

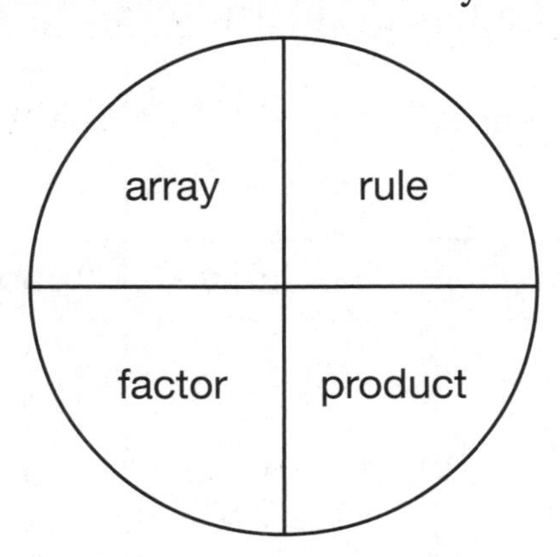

Strategy Focus

Look for a Pattern

MATH FOCUS: Multiplication Concepts

Learn About It

Read the Problem

So many ladybugs! Naomi can catch as many as she wants. On Monday, she caught 4 ladybugs. On Tuesday, she caught 8 ladybugs. On Wednesday, she caught 12 ladybugs. On Thursday, she caught 16 ladybugs. If this pattern continues, how many bugs will Naomi catch on Saturday?

Reread As you read the problem again, ask yourself these questions.

- What is the main idea of the problem?

____

- What facts do I know?

____

- What does the problem ask me to find?

____

Search for Information

Read the problem again. Circle the numbers you need.

Record Next, make a table to show the days and the number of ladybugs.

Monday	Tuesday	Wednesday	Thursday
• • • •	• • • • • • • •	• • • • • • • • • • • •	• • • • • • • • • • • • • • • •
4	8		

Check to be sure you wrote the correct numbers in the table.

Decide What to Do

You know how many ladybugs Naomi caught each day. You know that the number increases each day. You also know that the pattern continues.

Ask How many ladybugs will Naomi catch on Saturday?

- I can use the strategy *Look for a Pattern.*
- I can find how many more ladybugs Naomi catches each day.
- Once I find the pattern, I can add to find the number of bugs Naomi will catch on Friday and Saturday.

Use Your Ideas

Step 1 Look at the numbers in the table: 4, 8, 12, 16.

What do you add to 4 to get 8? 4 + ________ = 8

What do you add to 8 to get 12? 8 + ________ = 12

What do you add to 12 to get 16? 12 + ________ = 16

Each day, the number of bugs goes up by ________.
The rule for the pattern is *add* ________.

Step 2 Add to continue the pattern.

Monday	Tuesday	Wednesday	Thursday	Friday	Saturday
4	8	12	16		

+ 4 + 4 + 4 ________ ________

You can continue the pattern to solve the problem.

Naomi will catch ________ ladybugs on Saturday.

Review Your Work

Be sure your answer is for the correct day—Saturday.

Describe What did you do to find the pattern?

____

____

Try It

Solve the problem.

① Cara got a kitten named Tom. Cara had owned Tom for 7 days after 1 week, 14 days after 2 weeks, 21 days after 3 weeks, and 28 days after 4 weeks. How many days will Cara have owned Tom after 6 weeks?

Read the Problem and Search for Information

Identify the numbers and words you need in the problem. Circle them.

Decide What to Do and Use Your Ideas

You can use the strategy *Look for a Pattern* to answer the question. Make a table.

Week	1	2	3	4	5	6
Days	7	14	21	28		

Decide which operation you can use to see if the numbers make a pattern.

Ask Yourself

How can I get the number of days after 5 weeks?

Step 1 Write a rule for the pattern.

Add ________ to each number of days to get the next number of days.

Step 2 Use the rule to find the number of days in 5 weeks and then in 6 weeks.

After 5 weeks, Cara will have owned Tom for ________ days.

So after 6 weeks, Cara will have owned Tom for ________ days.

Review Your Work

Check that your addition was correct.

Illustrate How could you use dots to make an array to answer Problem 1?

____

____

Apply Your Skills

Solve the problems.

② Each month Pablo adds goldfish to his backyard pond. In April, Pablo adds 3 goldfish. In May, he adds 6 goldfish. In June, he adds 9 goldfish. In July, he adds 12 goldfish. If this pattern continues, in what month will Pablo add 18 goldfish?

Month	April	May	June	July		
Goldfish	3	6	9	12		

Hint Complete the table. You can draw dots to help you find the pattern.

Ask Yourself

What is the rule?

Answer ______________________________

Relate How can making a table help you find a pattern?

____

____

③ Lynn is buying fish. The table below shows the cost of the fish. If Lynn has $15, does she have enough money to buy 6 fish? How do you know?

Number of Fish	1	2	3	4	5	6
Cost	$2	$4	$6	$8		

Hint The cost increases by the same amount each time.

Ask Yourself

How can I find how much 5 fish cost?

Answer ______________________________

Generalize What is a rule for this pattern? Tell how you know.

____

____

Ask Yourself

How can I organize the information to look for a pattern?

Hint Find the number you add each time.

④ Tina takes her puppy for a walk every day. The table shows how long she walks her puppy each day. Suppose the time increases each day in the same way. How many minutes will she walk her dog on Saturday?

Sun.	Mon.	Tue.	Wed.			
5 min	10 min	15 min	20 min			

Answer ______________________________

Explain How can you use skip counting to check your work?

____

____

Ask Yourself

Which operations can I use to solve this problem?

⑤ Manuel wants to find out how many sheep live at his uncle's farm. In the barn, he counts 4 sheep in each of the 6 pens. He counts 3 more sheep in the field. How many sheep does he count altogether?

Pens	1	2				
Sheep	4	8				

Hint Do not forget to add the sheep in the field.

Answer ______________________________

Sequence What steps did you use to solve this problem?

____

____

On Your Own

Solve the problems. Show your work.

(6) Toshi spends $8 to feed his pet rabbit for 1 week. He spends $16 to feed his rabbit for 2 weeks. He spends $24 for 3 weeks and $32 for 4 weeks. This pattern continues. How much will Toshi spend to feed his rabbit for 6 weeks?

Answer ______________________________

Organize How did you show the information so you could look for a pattern?

____

____

(7) Ali uses string and paper turtles to make chains. She puts 3 turtles on the first chain. She puts 6 turtles on the second chain. For the third chain, she uses 9 turtles. For the fourth chain, she uses 12 turtles. If Ali continues this pattern, how many turtles will she use on the seventh chain?

Answer ______________________________

Adapt How could you use your work to find the total number of turtles Ali uses to make all 7 chains?

____

____

Create

Look at Problem 4. Change the number of days. Change the number of minutes Tina walks her dog each day. Write and solve your new problem.

Strategy Focus

Write a Number Sentence

MATH FOCUS: Multiplying Greater Numbers

Learn About It

Read the Problem

Gorillas and chimpanzees are both apes. Chimpanzees are often called "chimps" for short. Gorillas are larger than chimps. A chimp in a wildlife park weighs 110 pounds. A gorilla in the park weighs 4 times as much. How much does the gorilla in the park weigh?

Reread Read the problem again. Answer these questions.

- What is the problem about?

____

- Which weighs more, the gorilla or the chimp?

____

- What am I asked to find?

____

Search for Information

Read the problem again. Look for information you need.

Record What facts can you use to find the answer?

The chimp in the wildlife park weighs ________ pounds.

The gorilla weighs ________ times as much as the chimp.

These facts can help you choose a strategy.

Decide What to Do

You know how much the chimp weighs. You also know the gorilla weighs 4 times as much as the chimp.

Ask How can I find how much the gorilla in the park weighs?

- I can use the strategy *Write a Number Sentence.*
- I can write a multiplication sentence.

Use Your Ideas

Step 1 To write a multiplication sentence, decide whether each number is a factor or a product.

- The weight of the chimp is a *factor*.
- The gorilla weighs 4 times as much as the chimp. That number of times is also a *factor.*
- The weight of the gorilla is a *product*.

Step 2 Use what you know to write a multiplication sentence in words. Multiply to find the weight of the gorilla.

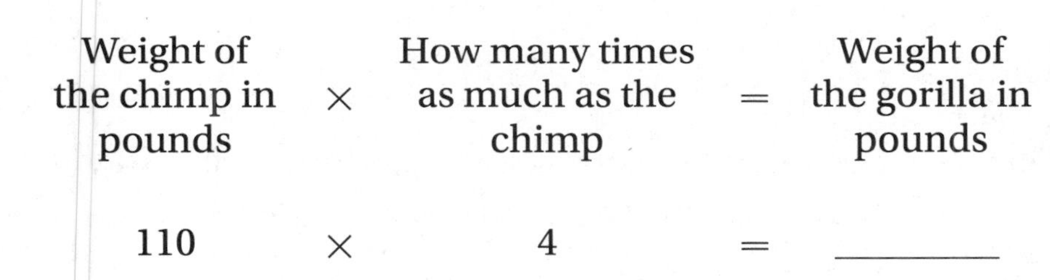

Weight of the chimp in pounds	×	How many times as much as the chimp	=	Weight of the gorilla in pounds
110	×	4	=	______

You can also multiply like this:

$$\begin{array}{r} 110 \\ \times \quad 4 \\ \hline \end{array}$$

So the gorilla in the park weighs ______ pounds.

Review Your Work

Look back at the problem. Be sure your answer makes sense.

State How can you tell that the weight of the gorilla is the product of the other two numbers of the problem?

____

____

Try It

Solve the problem.

1. Ms. Lee studies green sea turtles. On Monday, she found 2 turtle nests. There were 62 eggs in each nest. On Tuesday, she found 98 turtle eggs. Did Ms. Lee find more eggs on Monday or Tuesday? How many more?

Read the Problem and Search for Information

Retell the problem in your own words. Identify the facts.

Decide What to Do and Use Your Ideas

You can write a number sentence. You can use the number sentence to find how many eggs Ms. Lee found on Monday. Then you can compare the numbers of eggs for each day.

Step 1 Write a multiplication sentence in words. Multiply to find the number of eggs Ms. Lee found on Monday.

Number of nests	×	Number of eggs in each nest	=	Number of eggs found on Monday
2	×	________	=	________

Ask Yourself

Why should I subtract to find the answer to the second question?

Step 2 Compare the numbers of eggs found each day. Then subtract to answer to the second question.

Ms. Lee found ________ eggs on Monday.

Ms. Lee found ________ eggs on Tuesday.

________ − ________ = ________

Ms. Lee found more eggs on ________________ .

She found ________ more eggs.

Review Your Work

Check that you solved the number sentences correctly.

Recognize Which operation did you use to find the total number of eggs on Monday? Why?

____

____

Apply Your Skills

Solve the problems.

(2) The python is one of the longest snakes in the world. Pita made a model of a python. Her model is 3 feet long. A real python can grow to be about 10 times as long as her model. About how many feet long can a real python be?

Hint The length of a real python is a product of two numbers.

Ask Yourself

Which numbers in the problem should be the factors?

Length of Pita's model in feet	×	Number of times as long as Pita's model	=	Length of a real python in feet
______	×	10	=	______

Answer ____________________

Explain Seth says that the length of a real python is 10 feet. What might his error have been?

____

____

(3) A seal at an aquarium weighs 255 pounds. It eats 13 pounds of fish each day. How many pounds of fish does the seal eat in a week?

Ask Yourself

How many days are in a week?

Pounds eaten each day	×	Number of days in a week	=	Pounds eaten in a week
______	×	______	=	______

Hint One of the factors is a number given in the problem.

Answer ____________________

Identify What information is *not* needed to solve the problem?

____

____

④ A gray bat has a wingspan of about 30 centimeters. A flying fox bat has a wingspan 6 times as long as the wingspan of a gray bat. How much longer is the wingspan of a flying fox bat?

Hint Multiply to find the wingspan of a flying fox bat.

Wingspan of a gray bat in centimeters	×	Number of times as long as wingspan of a gray bat	=	Wingspan of a ______ in centimeters
______	×	______	=	______

Ask Yourself

What do I need to do after I multiply?

Answer ______________________________

Compare How is this problem like another problem in this lesson?

____

____

⑤ Dr. Jean is an animal doctor. She counts the teeth of 5 horses and 6 sheep at Sunset Ranch. How many horse teeth and how many sheep teeth does Dr. Jean count?

Farm Animal	Number of Teeth for Each Animal
Rabbit	28
Horse	42
Pig	44
Sheep	32

Hint You need two number sentences. One is for horse teeth. The other is for sheep teeth.

Horse Teeth: ______ × ______ = ______

Sheep Teeth: ______ × ______ = ______

Ask Yourself

What factors should I use in each number sentence?

Answer ______________________________

Decide What is another question the problem could ask?

____

____

On Your Own

Solve the problems. Show your work.

6. A certain kind of turtle can swim 17 kilometers in 1 day. In 1 day, a humpback whale can swim 6 times as far as the turtle can. How far can a humpback whale swim in 1 day?

Answer ______________________________

Generalize How can writing a number sentence help you solve this problem?

____

____

7. Lina recorded how many insects she saw in July. She saw 63 bees. She saw twice as many fireflies as bees. She also saw 8 times as many ants as fireflies. How many ants did Lina see?

Answer ______________________________

Conclude Why do you need to know the meaning of the word *twice* to solve the problem?

____

____

Create

Look at Problem 3. Change the type of animal and how many pounds of food it eats each day. Write and solve your new problem.

Lesson 7

Strategy Focus
Act It Out

MATH FOCUS: Division Concepts

Learn About It

Read the Problem

Jared has 12 pictures of his trip to the lake. He plans to put them in his scrapbook. He wants to put an equal number of pictures on each page he uses. Jared can put up to 6 pictures on a page. How many pages could Jared use in his scrapbook?

Reread Ask yourself questions as you read.

- What is the problem about?

____

- What is the greatest number of pictures Jared can put on a page?

____

- What am I asked to find?

____

Search for Information

Circle numbers and words you may need.

Record What facts will help you solve this problem?

Jared has ________ pictures.

Jared wants to put an ________ number of pictures on each page.

No more than ________ pictures will fit on a page.

You can use these facts to help you think about how to solve the problem.

Decide What to Do

You know there are 12 pictures.

Ask How can I find how many pages Jared could use in his scrapbook?

- I can use the strategy *Act It Out*.
- I can use objects to find the ways 12 pictures can be separated into equal groups.

Use Your Ideas

Step 1 Use ________ connecting cubes to stand for Jared's pictures. Start by trying to separate the 12 connecting cubes into equal groups of 6.

Jared can put 6 pictures on each page. He will use ________ pages.

Step 2 Try to separate the 12 connecting cubes into equal groups of 5, 4, 3, 2, and 1.

12 connecting cubes cannot be divided into equal groups of 5.

Jared can put 6 pictures on a page. He will use 2 pages.

Jared can put 4 pictures on a page. He will use ________ pages.

Jared can put 3 pictures on a page. He will use ________ pages.

Jared can put 2 pictures on a page. He will use ________ pages.

Jared can put 1 picture on a page. He will use ________ pages.

Review Your Work

Check that the same number of pictures is on each page.

Explain Why is the strategy *Act It Out* useful for solving this problem?

____

____

Try It

Solve the problem.

① The table shows the lengths of different kinds of canoes that a store sells. How many times as long as the River Rider is the Fin?

Canoe	Length (feet)
Fin	18
White Wave	12
River Rider	9
Big Splash	20

Read the Problem and Search for Information

Reread the problem. Circle the numbers you need.

Decide What to Do and Use Your Ideas

You can use the strategy *Act It Out* to solve the problem.

Ask Yourself

How can I use the connecting cubes to solve the problem?

Step 1 Use connecting cubes to show the canoes.

Length of the Fin (18 feet)

Length of the River Rider (9 feet)

You need ________ cubes for the Fin.

You need ________ cubes for the River Rider.

Step 2 Separate the connecting cubes for the Fin into lengths that are as long as the River Rider.

There are ________ lengths of 9 in 18.

So the Fin is ________ times as long as the River Rider.

Review Your Work

Check that you answered the question that was asked.

Identify What information is given in this problem that you do *not* need to use?

____

____

Apply Your Skills

Solve the problems.

(2) Mimi collects rocks at the beach. She uses the rocks to decorate her garden. One week, she collected 35 rocks. She collected the same number of rocks each day of the week. How many rocks did Mimi collect each day?

Ask Yourself

How many equal groups do I need to make?

Mimi collected ________ rocks in one week.

There are ________ days in one week.

I need to separate ________ connecting cubes into ________ equal groups.

Hint Make sure you put the same number of connecting cubes in each group.

Answer ____________________

Describe Why do you need to separate the connecting cubes into equal groups to solve the problem?

____

____

(3) Workers are making a walkway from the snack bar to the lake. The walkway will be 30 feet long. The workers have boards that are each 3 feet long. How many boards do the workers need to make the walkway?

Hint You can use connecting cubes to show the length of the walkway and the length of one board.

The length of the walkway is ________ feet.

The length of each board is ________ feet.

Ask Yourself

How many equal groups of 3 are in 30?

Answer ____________________

Model Describe how you could use connecting cubes to solve this problem.

____

____

Ask Yourself

What information in the table do I need?

Hint Circle the numbers in the table that you will use.

4. The table below shows the lengths of bridges over some rivers. How many times as long as Bridge B is Bridge D?

Bridge	Length (miles)
Bridge A	4
Bridge B	8
Bridge C	20
Bridge D	24

Bridge D is ________ miles long.

Bridge B is ________ miles long.

Answer ______________________________

Conclude Suppose Anton's answer was that Bridge D was 16 miles longer than Bridge B. What was Anton's error?

____

____

Hint The number of white shells on each frame will not be equal to the number of gray shells on each frame.

Ask Yourself

How can I use connecting cubes to show the two kinds of shells?

5. Josie uses shells to decorate 9 frames. She has 45 white shells and 36 gray shells. Each frame will have an equal number of white shells. Each frame will also have an equal number of gray shells. How many white shells will Josie use on each frame? How many gray shells?

There will be ________ frames.

Answer ______________________________

Determine How can you use multiplication to check your answers?

____

____

On Your Own

Solve the problems. Show your work.

6. Luis swims 36 meters across a lake. Kim swims 6 meters from the beach to a raft, then 6 meters back to the beach. How many times must Kim swim to the raft and back to go as far as Luis?

Answer ______________________________

Adapt What other question could you ask from the problem?

____

____

7. Every day for one week, 24 campers went on boat rides. They went in a different type of boat each day. The table shows the kinds of boats they used. It also shows how many campers one boat can hold. How many more paddleboats than rowboats did they use?

Day	Type of Boat	Campers in One Boat
Mon.	Canoe	3
Tue.	Kayak	2
Wed.	Paddleboat	2
Thu.	Rowboat	6
Fri.	Sailboat	4

Answer ______________________________

Sequence What steps did you take to solve this problem?

____

____

Look at Problem 3. Change the distance from the snack bar to the lake. Change the length of the boards. Write and solve your new problem.

Strategy Focus

Solve a Simpler Problem

MATH FOCUS: Dividing Greater Numbers

Learn About It

Read the Problem

Don grows tomatoes in his garden. He picks 127 tomatoes on Friday and 177 tomatoes on Saturday. Don puts the tomatoes in boxes to sell. Each box holds 4 tomatoes. How many boxes will he need to hold all the tomatoes?

Reread As you read the problem, ask yourself questions.

- On what days does Don pick tomatoes?

____

- What does Don do with the tomatoes he picks?

____

- What do I need to find?

____

Mark the Text

Search for Information

Read the problem again. Underline numbers you will need.

Record Look for facts that will help you solve the problem.

On Friday, Don picks ________ tomatoes.

On Saturday, Don picks ________ tomatoes.

Each box holds ________ tomatoes.

Think about how you can use this information to solve the problem.

Decide What to Do

You know how many tomatoes Don picked each day.

Ask How can I find the number of boxes Don needs?

- I can use the strategy *Solve a Simpler Problem.*
- I can use simpler numbers to decide how to solve the problem. Then I can use the numbers from the problem.

When the numbers are hard to work with or when there is more than one step, try solving a simpler problem first.

Use Your Ideas

Step 1 Choose simpler numbers that are easy to work with. Try 8 for Friday and 12 for Saturday.

Step 2 Use the numbers to see how to solve the problem.

Add to find the number of tomatoes picked in all.

$8 + 12 =$ ________

Divide to find the number of boxes needed.

$20 \div 4 =$ ________

Five boxes are needed.

Step 3 Use your plan to solve the original problem.

Add to find the number of tomatoes picked in all.

________ + ________ = ________

Divide to find the number of boxes needed.

________ ÷ ________ = ________

So Don will need ________ boxes to hold all the tomatoes.

Review Your Work

Check that you followed your plan.

Summarize How did you use the strategy *Solve a Simpler Problem* to solve this problem?

____

____

Try It

Solve the problem.

① Maya, Paul, and Lee pick 235 berries. Altogether they eat 28 berries. Then they share the rest of the berries equally to bring home. How many berries does Maya bring home?

Read the Problem and Search for Information

Think about what the friends do with the berries.

Decide What to Do and Use Your Ideas

You can use the strategy *Solve a Simpler Problem.*

Step 1 Choose simpler numbers that are easy to work with.

	Number in Problem	Simpler Number
Number of berries picked		23
Number of berries eaten		2

Step 2 Use the numbers to plan how to solve the problem.

- First, I will ____________________ .

  $23 - 2 =$ _________

- Then I will ____________________ .

  _________ $\div 3 =$ _________

Which operation can I use to show that the friends ate some berries?

Step 3 Use your plan to solve the original problem.

Subtract. _________ $-$ _________ $=$ _________

Divide. _________ $\div$ _________ $=$ _________

So Maya brings home _________ berries.

Review Your Work

Check that you followed your plan correctly.

Conclude How does using 23 and 2 instead of 235 and 28 help you solve this problem?

____

____

Apply Your Skills

Solve the problems.

② Mr. Dunn puts plants into trays to sell. He has 128 mint plants and 176 pepper plants. How many trays can he fill with mint plants? How many can he fill with pepper plants?

Hint Use information from the picture to help you solve the problem.

Choose simpler numbers that are easy to work with. Make a plan.

Suppose there are 6 plants and each tray holds 2.

Find how many trays can be filled.

6 ◯ 2 = ________

Then follow your plan using the numbers in the problem.

Ask Yourself

Which operation will I use to find the number of trays Mr. Dunn can fill with the plants?

Answer ______________________________

Examine How did the picture help you solve this problem?

____

____

③ A nature club plants a total of 224 trees. Of those trees, 128 are pine trees. The rest of the trees are maple trees. There are 6 people in the club. Each person plants the same number of maple trees. How many maple trees does each person plant?

Hint Subtract to find the number of maple trees planted.

Number of trees planted in all: ________

Number of pine trees planted: ________

Number of maple trees planted: ________

Ask Yourself

What operation can I use to find the number of maple trees each person plants?

Answer ______________________________

State What steps did you use to solve the problem?

____

____

Ask Yourself

What simpler numbers can I use?

4 Ms. Roth has 241 red tulips, 302 yellow tulips, and 117 pink tulips. She puts the flowers in groups of 6. How many groups of tulips does she make?

Hint Look at the numbers of tulips. Use the first digit in each number as a simpler number.

Add the simpler numbers.

_______ + _______ + _______ = _______

Find the number of groups.

_______ ÷ _______ = _______

Now follow your plan using the numbers in the problem.

Answer ___________________________

Determine How can you estimate to check that your answer is reasonable?

____

____

5 The local garden center sells plants. The table shows how much each plant costs. The table also shows the sales for each type of plant. How many more daisies did the garden center sell than roses?

Hint The words *how many more daisies* tell you that more daisies were sold than roses.

Plant	Cost	Sales
Daisies	$3	$156
Mums	$5	$415
Sunflowers	$4	$176
Roses	$7	$203

Ask Yourself

Which operation can I use to find the number of daisies that were sold?

How many daisies were sold? _______

How many roses were sold? _______

Answer ___________________________

Judge What information is *not* necessary to answer the question the problem asks?

____

____

On Your Own

Solve the problems. Show your work.

6. Jodi and Lucy picked 136 apples altogether and put them in bags. They put 8 apples in each bag. Jodi filled 5 bags. How many bags did Lucy fill?

Answer ______________________

Consider Would your plan change if the problem said Jodi and Lucy picked 56 apples instead of 136? Explain.

____

____

7. Mrs. Lee is planting sunflowers. Last year she planted 135 sunflowers. This year, she plants 3 times as many sunflowers. Each row she plants has 9 sunflowers. How many rows of sunflowers are there?

Answer ______________________

Represent Using simpler numbers, draw a picture of the problem and show how to solve it.

____

____

Create

Look at Problem 3. Write a new problem. Change the number of people in the club and the number of trees they plant. Then solve your problem.

UNIT 2 Review What You Learned

In this unit, you worked with four problem-solving strategies. You can often use more than one strategy to solve a problem. So if a strategy does not seem to be working, try a different one.

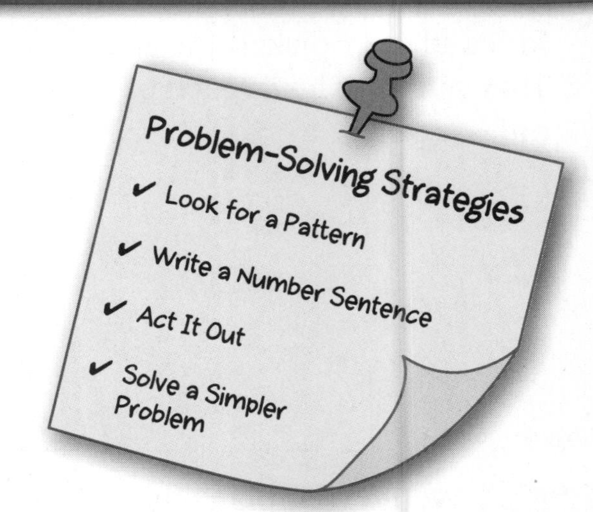

Solve each problem. Show your work. Record the strategy you use.

1. Mr. Carter works at the market. He puts 168 boxes of cereal on shelves. He fills 8 shelves. He puts an equal number of boxes on each shelf. How many boxes are on each shelf?

Answer ____________________

Strategy ____________________

2. Mrs. Russo is making fruit salad for a party. She uses the recipe shown. If she makes the salad for 4 people, how many pieces of fruit will she need in all?

Fruit Salad Recipe (serves 1 person)

- 1 banana
- 3 strawberries
- 6 cherries

Answer ____________________

Strategy ____________________

3. A pencil factory makes a batch of 384 pencils. The pencils fill 6 boxes. Each box holds the same number of pencils. How many pencils are in each box?

Answer ______________________________

Strategy ______________________________

4. Kim is making 4 bracelets. She has 32 beads in all. Kim has 12 red beads and 20 green beads. Each bracelet has an equal number of red beads. Each bracelet also has an equal number of green beads. How many red beads and how many green beads are on each bracelet?

Answer ______________________________

Strategy ______________________________

5. Jim is buying tickets to the talent show. The ticket prices for more than 4 tickets follow the same pattern. Jim and 6 friends want to go to the talent show. How much will their tickets cost?

Park Street School Talent Show

Number of Tickets	Price
1	$3
2	$6
3	$9
4	$12

Answer ______________________________

Strategy ______________________________

Explain how you found the pattern.

Solve each problem. Show your work. Record the strategy you use.

6. Henry skip counts aloud by 3s. He uses the chart below. He shades each number he says. He stops at 27. If Henry keeps skip counting by 3s, what are the other numbers he should say to complete the chart? Write the numbers in order.

1	2	3	4	5	6	7	8	9
10	11	12	13	14	15	16	17	18
19	20	21	22	23	24	25	26	27
28	29	30	31	32	33	34	35	36
37	38	39	40	41	42	43	44	45

Answer ______________________________

Strategy ______________________________

7. Lori decorates tables at a party with groups of balloons. Each group of balloons has 7 yellow balloons and 3 white balloons. She has 245 yellow balloons and 105 white balloons. How many groups of balloons does Lori make?

Answer ______________________________

Strategy ______________________________

8. The pictograph shows the number of different kinds of fish at the aquarium. How many jellyfish are there?

Fish at the Aquarium

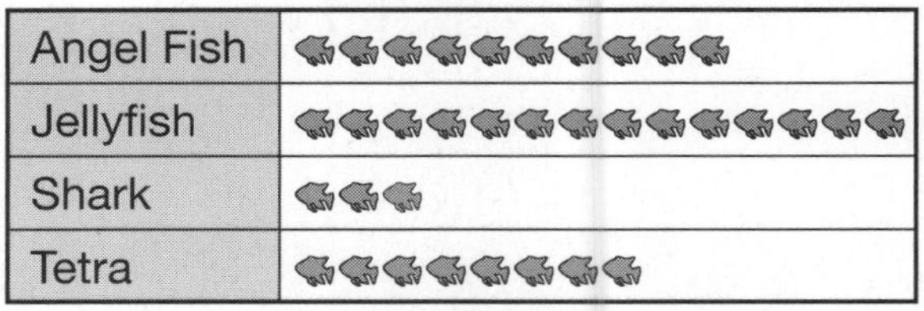

Angel Fish	
Jellyfish	
Shark	
Tetra	

Key: Each stands for 5 fish.

Answer ______________________________

Strategy ______________________________

Explain how you can skip count to solve the problem.

9. For a school party, Mr. Torres orders 15 boxes of blueberry muffins and 18 boxes of corn muffins. There are 12 muffins in each box. Does Mr. Torres have enough muffins for 400 students to each have one muffin? If not, how many more muffins does he need?

Answer ______________________

Strategy ______________________

10. The table shows how long Lola practices piano each day. If the pattern continues, on what day will Lola practice for 1 hour?

Day	1	2	3	4	5	6	7
Number of Minutes	10	20	30	40			

Answer ______________________

Strategy ______________________

Write About It

Look back at Problem 1. Explain how you can estimate to check that your answer is reasonable.

____

____

Work Together: Plan a Trip

Your group is planning a trip to a theme park for 48 students. A train will take the students around the park. You need to decide how many train cars to use.

Plan Find all the different ways to group the students in train cars that follow the rules.

Decide As a group, decide how many train cars you will use.

Draw Make a picture of your train. Show the people in each train car. Write how many stickers, postcards, and bags are needed for each car.

Present As a group, share your decision and picture with the class. Explain how you solved the problem.

Information About the Theme Park Train Ride

- The same number of students must be in each train car.
- No more than 8 students are allowed in a train car.
- Up to 12 cars can be in a train.

On the train, each student gets:

- 25 stickers
- 3 postcards
- 1 theme park bag

UNIT 3 Problem Solving Using Data, Graphing, and Probability

Unit Theme:
Having Fun

How do you have fun? You might spend your free time listening to music or riding a bike. Some people play computer games or go roller skating. In this unit, you will see how math plays a big part in having fun.

Math to Know

In this unit, you will use these math skills:

- Organize data
- Use line plots, pictographs, and bar graphs
- Understand probability and outcomes

Problem-Solving Strategies

- Make a Table
- Make a Graph
- Make an Organized List

Link to the Theme

Finish the story. What does Jeri tell her uncle? Include some of the facts from the table at the right.

Jeri's uncle owns a toy store. She is helping him count how many toys he has. He will order more toys if he is running out.

Toy Name	Number of Toys in Store
Panda Pals	18
Telly the Tractor	5
Travel Checkers	2

____

____

____

____

____

____

____

Use Math Language

Review Vocabulary

Here are some math words for this unit. Knowing what these words mean will help you understand the problems.

bar graph	line plot	pictograph	row
column	outcome	probability	tally chart

Vocabulary Activity   Multiple-Meaning Words

Some words mean one thing in math and something else in everyday English. Use words from the above list to complete the following sentences.

1. You can ________________ a boat around a pond.
2. A ________________ in a table goes from left to right.
3. A ________________ is shaped like a pole and holds up a roof.
4. A ________________ in a table goes from top to bottom.

Graphic Organizer   Word Circle

Complete the graphic organizer.

- Cross out the word that does *not* belong.
- Replace it with a word from the vocabulary list that does belong.

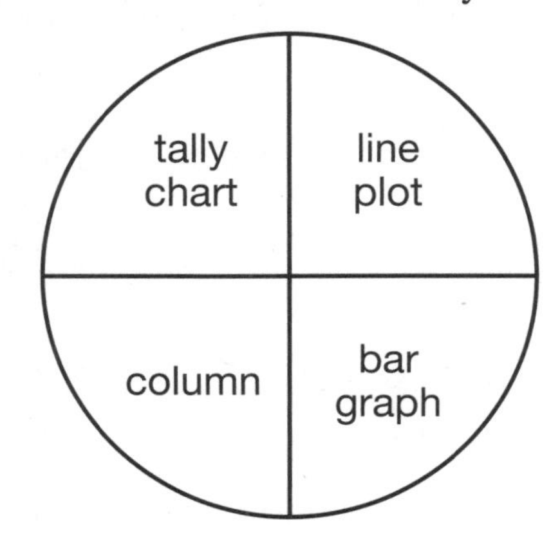

Strategy Focus

Make a Table

MATH FOCUS: Organize Data

Learn About It

Read the Problem

Each worker at the toy factory is supposed to paint at least 4 slimy snakes in 1 minute. A worker gets 10¢ more for each extra snake painted in 1 minute. In 1 minute, eight workers painted 4, 6, 3, 5, 7, 3, 3, and 3 snakes. How much does the toy factory need to pay for the extra snakes?

Reread Ask yourself these questions as you read.

- What is this problem about?

____

- What data are given?

____

- What question should I answer?

____

Mark the Text

Search for Information

Read the problem again. Circle numbers that you will need. Underline the question.

Record Write what will help you solve the problem.

The workers are supposed to paint at least ________ slimy snakes in 1 minute.

The toy company pays ________ more for each extra snake painted in 1 minute.

Think about how you will organize all the numbers that are given.

Decide What to Do

You know that workers are paid more for painting extra snakes. You know how many snakes eight workers painted.

Ask How can I find out how much more the toy factory needs to pay for the extra snakes?

- I can use the strategy *Make a Table* to organize the numbers I know.
- I can find how many extra snakes are painted. Then I can multiply that number by 10¢.

Use Your Ideas

Step 1 Make a table. Fill in the numbers of slimy snakes painted.

Step 2 Subtract to find the number of extra snakes for each worker.

Step 3 Add to find the total number of extra snakes. Then multiply 10¢ by that number.

Number of Snakes Painted	Number of Extra Snakes Painted
4	0
6	6 − 4 = 2
3	0
5	5 − 4 =

Add. 2 + 1 + ________ = ________

Multiply. ________ × 10¢ = ________ ¢

So the toy factory needs to pay the workers ________ ¢ extra.

The number of extra snakes is the number of snakes greater than 4 that a worker painted. Remember to subtract 4 from numbers 5 or greater to find the number of extra snakes for each worker.

Review Your Work

Make sure you answered the question that was asked.

Generalize How did a table help you solve the problem?

____

____

Try It

Solve the problem.

① The toy factory also makes dolls, trucks, and robots. This tally chart shows how many of each toy was made so far this week. How many more robots than dolls were made so far this week?

Toy	Mon.	Tues.	Wed.
Dolls	𝍸 𝍷𝍷	𝍸 𝍸	𝍸 𝍷𝍷
Trucks	𝍸 𝍷	𝍸 𝍷𝍷𝍷𝍷	𝍸 𝍷𝍷𝍷
Robots	𝍸 𝍷𝍷𝍷	𝍸 𝍷𝍷𝍷𝍷	𝍸 𝍷𝍷𝍷𝍷

Read the Problem and Search for Information

Look at the tally chart and read the problem again. Underline the question you need to answer.

Decide What to Do and Use Your Ideas

Make a table to record the data from the tally chart.

Ask Yourself

Do I need a row for trucks in my table?

Step 1 Make rows for dolls and for robots. Make columns for Monday, Tuesday, Wednesday, and the Total.

Toy	Mon.	Tue.	Wed.	Total
Dolls	7	10	7	24
Robots				

Step 2 Fill in the table with the number of toys made. Add to find the total for each toy. Subtract to find how many more robots than dolls were made.

So there were ________ more robots than dolls made.

Review Your Work

Check that you entered the correct total for each toy.

Identify Which words helped you know to subtract?

____

____

Apply Your Skills

Solve the problems.

② Ms. Garcia sold all the toys on her sale table at her toy store. She sold 2 drums, a baseball, 3 dolls, a pony, and 3 MP3 players. The MP3 players sold for $12 each. The other toys sold for $8 each. What was the total amount that people paid for those items?

	Number	Price	Amount
MP3 Players		$12	
Other Toys		$8	
		Total Amount	

Answer ____________________

Recognize How did you figure out what to write for the number of "Other Toys" in the table?

____

____

Ask Yourself

How many baseballs did Ms. Garcia sell? How many ponies?

Hint You need to multiply and then add to find the total amount paid.

③ Ms. Goldberg sells stuffed toys at her store. The stuffed toys are 3 kinds of animals in different colors. She has 2 blue cats, 6 green dogs, 1 pink cat, 3 pink pigs, 4 purple cats, 4 purple dogs, and 2 green pigs. Ms. Goldberg wants 10 of each animal. How many more of each animal should she order?

	Total	Number to Order
Cats	7	
Dogs		
Pigs		

Answer ____________________

Describe Which math operations did you use to solve this problem? Describe how you used them.

____

____

Hint You need to think about the animals, not their colors. Cross out the names of the colors to help you see the information you need.

Ask Yourself

How can I use a table to help find the answer?

Ask Yourself

How many of each item did D.J.'s Sports Store sell? What was the price of each item?

Hint Multiply to find the amount for each item. Then add to find the total amount that people paid.

4. Yesterday morning, D.J.'s Sports Store sold 3 sports caps for $25 each. It sold 4 soccer balls for $8 each. It sold 5 T-shirts for $10 each. What was the total amount that people paid for those items?

Item	Number	Price	Amount
		Total Amount	

Answer ____________________

Compare How is this problem like Problem 2?

____

____

5. Mr. Obi lost the list of things he ordered for his toy store. He remembers there were 6 games for $5 each. There were 8 baseballs for $4 each. There were some craft kits for $2 each. Mr. Obi paid $72 for his order. How many craft kits were in the order?

Hint Use the table to find the total amount he paid for the games and baseballs.

		Total Amount	

Mr. Obi paid ________ in all.

Mr. Obi paid ________ for the craft kits he ordered.

Ask Yourself

If I know the total amount Mr. Obi paid for his order, how can I find how much he paid for all the craft kits?

Answer ____________________

Explain How can you check your answer?

____

____

On Your Own

Solve the problems. Show your work.

6. A toy company makes toy cars. The toy company puts the cars in boxes of two different sizes. The toy company has to buy the boxes. Boxes for toy cars shorter than 6 inches cost $1. Boxes for toy cars 6 inches long or longer cost $2. The toy company makes five 12-inch cars, four 10-inch cars, and eight 4-inch cars. How much will the boxes for those toy cars cost?

Answer ______________________________

Formulate What is another question you could ask about the data in this problem?

____

____

7. Mr. Vincent eats lunch in the company lunch room. A full-price lunch costs $5.00. For every full-price lunch he buys, he gets $1 off on his next lunch. Then on the lunch after that, he pays full price again. He buys 5 lunches this week. How much does Mr. Vincent spend on lunch this week?

Answer ______________________________

Elaborate Suppose Mr. Vincent buys lunch 10 days in a row. How much does he spend on lunch for those 10 days? Explain.

____

____

Create

Look back at the problems in this lesson. Choose one problem and change at least two numbers to create a new problem. Solve your new problem.

Strategy Focus
Make a Graph

MATH FOCUS: Line Plots

Learn About It

Read the Problem

Bess and her friends were playing a game with hula hoops. Bess counted their spins. She made this table to keep track. Which number of spins happened most often?

Hula-Hoop Spins									
Bess	Kee	Ruth	Drew	Abu	Lori	Luz	Jack	Nita	Tom
4	8	9	9	7	10	8	7	8	15

Reread Find the information you need.

- What did Bess make?

____

- Why did Bess make a table?

____

- What question do you need to answer?

____

Search for Information

Read the problem again. Underline the question you need to answer. Then look at the table.

Record Write the data you can find from the table.

The number of friends who were playing a game was ________ .

The table shows the number of ________________ for each friend.

Think about a different way to organize all the numbers.

Decide What to Do

You have a table that shows the number of spins for each friend.

Ask How can I find which number of spins happened most often?

- I can use the strategy *Make a Graph.*
- I can make a line plot to compare the numbers.

Use Your Ideas

Step 1 Make a line plot to show the numbers in the table. (The first two numbers, 4 and 8, are on the line plot as examples.) Write *Number of Spins* under the line plot to show what the numbers mean.

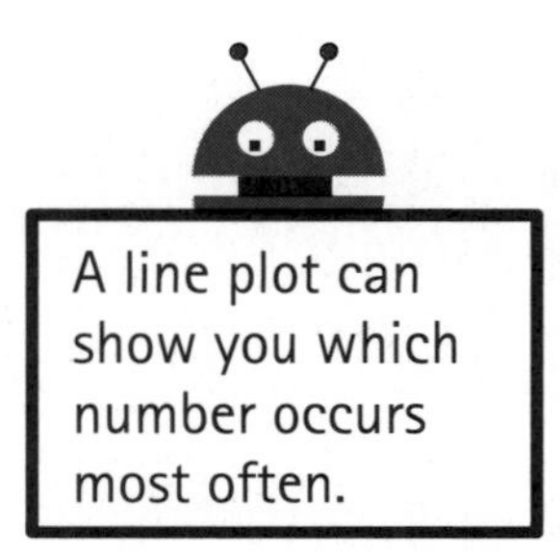

X				X							
4	5	6	7	8	9	10	11	12	13	14	15

Number of Spins

Step 2 Look for the number with the most Xs. That number is ________ .

So the number of spins that happened most often is ________ .

Review Your Work

Check that the number of Xs on your line plot matches the number of friends playing the game.

Describe How does making a line plot help you find the answer quickly?

____

____

Try It

Solve the problem.

① Chang has 18 marbles. He wonders if any students in his school have more than 20 marbles. So Chang asks 25 students how many marbles they have. He made this list. How many students have more than 20 marbles?

12, 24, 23, 16, 16, 14,
19, 19, 15, 13, 23, 26,
27, 16, 17, 26, 28, 29,
24, 15, 13, 17, 13,
16, 19

Read the Problem and Search for Information

Read the problem again. Think about how you can find the number of students who have more than 20 marbles.

Decide What to Do and Use Your Ideas

You can make a graph to count the numbers.

Step 1 Start a line plot. Draw and label a number line from 12 through 29. Plot the numbers in Chang's list.

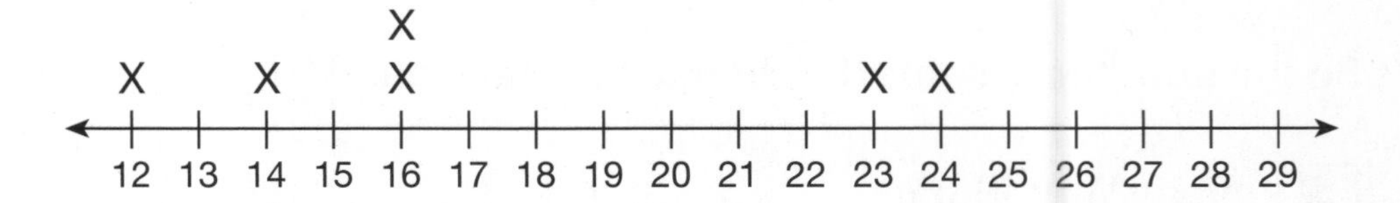

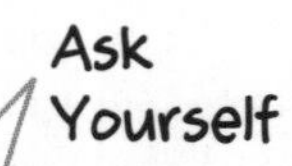

Do I look at all the Xs on the line plot or only the numbers to the right of 20?

Step 2 Count the Xs for numbers greater than 20.
There are ________ numbers greater than 20.

So ________ students have more than 20 marbles.

Review Your Work

Check that you made an X for each student.

Distinguish Which number is *not* needed to find the answer?

____

____

Apply Your Skills

Solve the problems.

2 Dee's class went for a nature walk. The students made the list below to record how many squirrels each saw. Which number of squirrels did the most students see?

7, 5, 8, 3, 9, 5, 6, 3, 2, 6, 4, 5, 7, 8, 5, 6, 6, 7, 4, 6, 3, 5, 6, 4

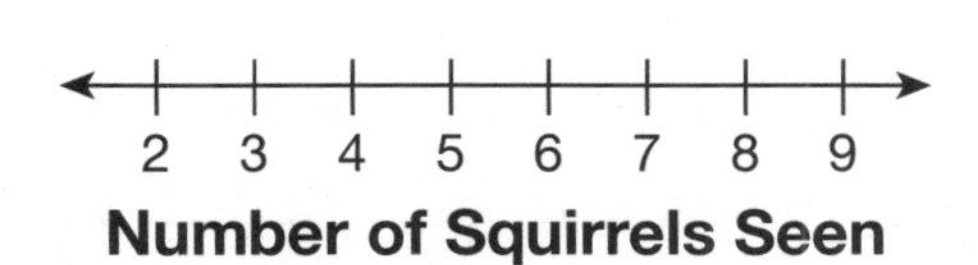

> **Hint** Make a line plot. You can cross out the numbers in the list as you graph each one.

> **Ask Yourself**
> How many students went on the nature walk? Is that the same as the number of Xs on my line plot?

Answer ______________________________

Relate How can you check that your line plot shows all the numbers in the box?

____

____

3 Jay and his friends did jumping jacks. They made the list below to record how many jumping jacks they each did in 1 minute. Jay said more of his friends did 42 jumping jacks than any other number. Is Jay right? Tell how you decided.

42, 36, 40, 39, 40, 38, 40, 41, 42, 36, 42, 38, 40, 42, 41, 39, 40, 41

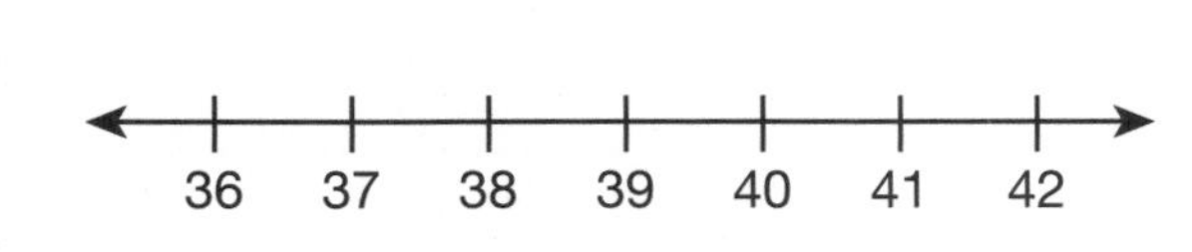

____

> **Hint** Make a line plot. Label your line plot to show what the numbers mean.

> **Ask Yourself**
> How many Xs should there be on my line plot?

Answer ______________________________

Sequence What steps did you take to solve this problem?

____

____

Hint Be sure to include each number in the list exactly once in your line plot.

Ask Yourself

What do I need to look for on the line plot after I have made all the Xs?

④ Rosa asks her friends to tell her their favorite number that is less than 15. She makes this list: 7, 3, 9, 13, 2, 5, 4, 2, 5, 4, 9, 7, 3, 1, 6, 8, 6, 8, 1, 7, 2, 8. What are the three most popular numbers?

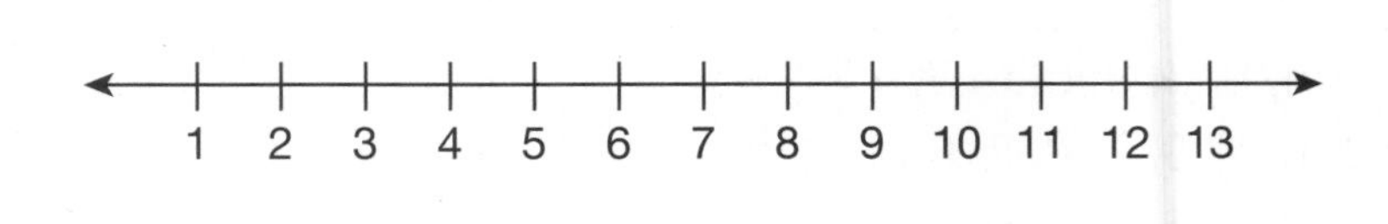

____

Answer ______________________

Identify Why isn't there an X above every number?

____

____

⑤ Mary has stilts. She can walk 17 steps on them. Some friends tried the stilts. Mary counted their steps. They walked 14, 18, 16, 24, 14, 13, 20, 19, 15, and 11 steps. Mary thinks her number of steps is the middle of the data. Is she right? Use the data to support your answer.

Hint Include Mary's number of steps on the line plot.

Ask Yourself

How many friends walked fewer steps than Mary? How many walked more steps than Mary?

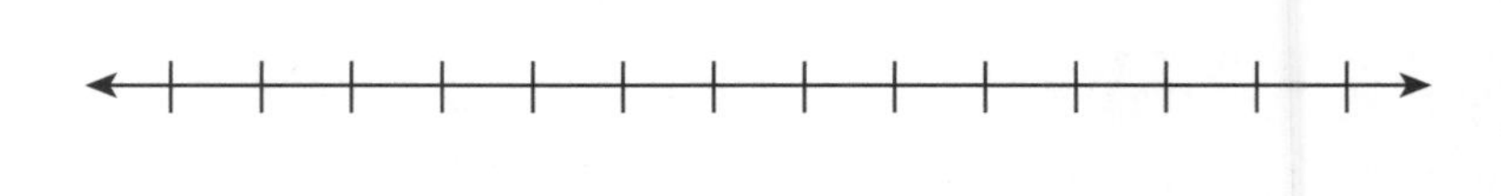

____

Answer ______________________

Develop What is another question that could be answered with the data in this problem?

____

____

On Your Own

Solve the problems. Show your work.

(6) Josie and her friends took turns using bubble mix. They counted the number of bubbles each person blew. Josie made a table to show the numbers. What number of bubbles was blown most often?

Number of Bubbles									
Josie	Bel	Cam	Faye	Nick	Lars	Jala	Adra	Bert	Gil
4	6	4	10	7	8	4	5	9	6

Answer ______________________________

Determine How is this problem like the Learn About It problem?

____

____

(7) Mr. Black set up a treasure hunt. His students have to find 20 items. The list below shows the number of items each student has found so far. Have more students found fewer than 12 or more than 12 items so far? Tell how you know.

10, 17, 13, 16, 14, 11, 10, 9, 8, 9, 14,
15, 13, 11, 13, 10, 11, 13, 16

Answer ______________________________

Consider What information is *not* needed to solve this problem?

____

____

Create

Choose a problem in this lesson. Change two numbers. Write and solve your new problem.

Lesson 11

Strategy Focus Make a Graph

MATH FOCUS: Pictographs and Bar Graphs

Learn About It

Read the Problem

Min is going to bring fruit for her class picnic. She asked the class to vote for their favorite fruit. Three students voted for pears. Seven students said they like bananas best. Three fewer students voted for bananas than apples. Two more students voted for oranges than pears. Which fruit was the most popular?

Reread Ask yourself these questions.

- What is the problem about?

____

- What am I asked to find?

____

Mark the Text

Search for Information

Reread the problem. Circle words that help you solve the problem.

Record Write what you know about how the students voted.

The number of students who voted for pears was ________ .

The number of students who voted for bananas was ________ .

Three fewer students voted for ________________ than apples.

Two more students voted for ________________ than pears.

Think about how you can show the data to solve the problem.

Decide What to Do

You know the kinds of fruit that the students in Min's class like. You also know how many students voted for pears and bananas.

Ask How can I tell which fruit is the most popular?

- I can use the strategy *Make a Graph.*
- I can make a bar graph to see which fruit was the most popular.

Use Your Ideas

Step 1 Make a bar graph.

Draw bars for the numbers you know.

Draw a bar to show ________ pears.

Draw a bar to show ________ bananas.

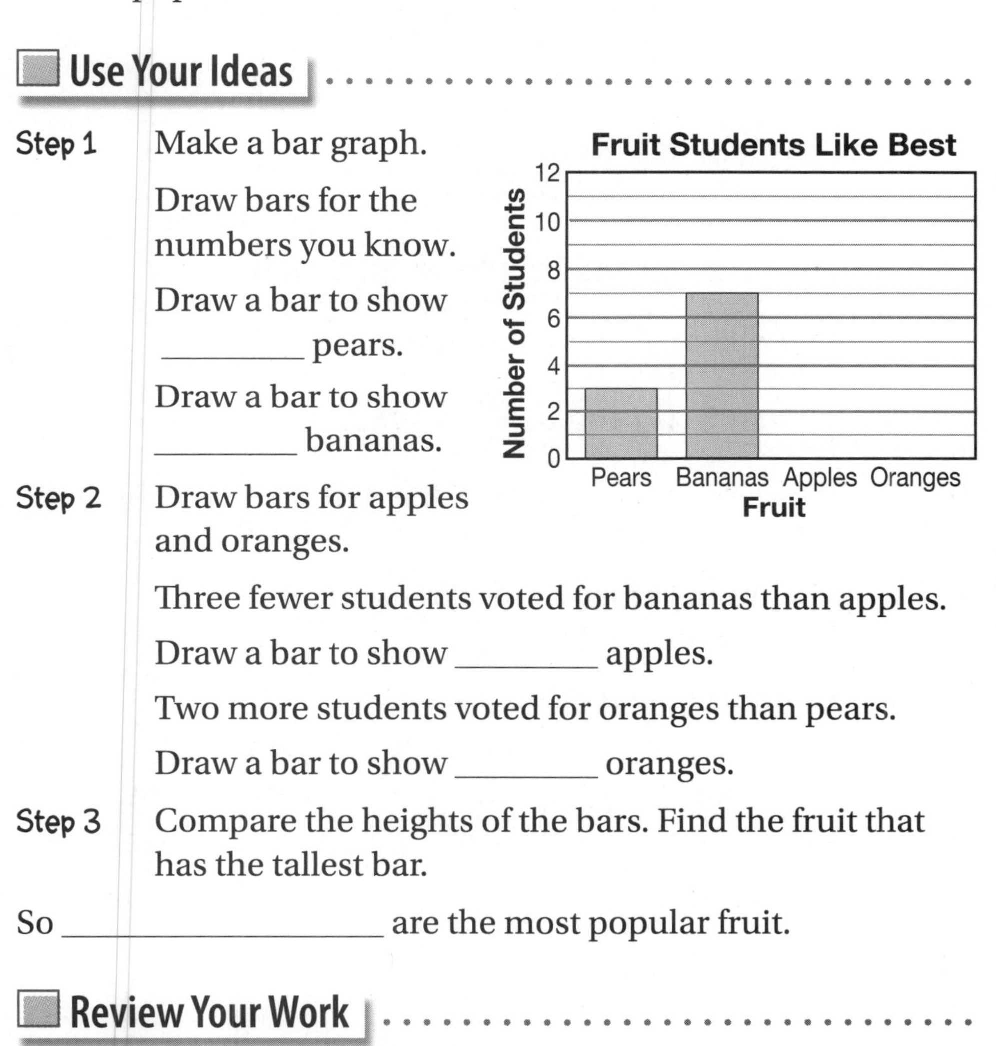

Step 2 Draw bars for apples and oranges.

Three fewer students voted for bananas than apples.

Draw a bar to show ________ apples.

Two more students voted for oranges than pears.

Draw a bar to show ________ oranges.

Step 3 Compare the heights of the bars. Find the fruit that has the tallest bar.

So ________________ are the most popular fruit.

You know the number for bananas. You can add to find how many students voted for apples.

Review Your Work

Check that each bar you drew shows the correct number.

Explain How does making a bar graph help you solve the problem?

____

____

Try It

Solve the problem.

① Ms. Jordan made a pictograph of the 3 kinds of bagels she sells in her bakery. It shows the number of blueberry bagels sold. It also shows that 100 plain bagels and 50 raisin bagels were sold. Peggy saw the pictograph. She said more than half the bagels sold were blueberry. Is Peggy correct? Explain how you know.

Number of Bagels Sold

Blueberry	○ ○ ○ ○ ○
Plain	
Raisin	

Key: Each ○ stands for 25 bagels.

Read the Problem and Search for Information

Reread the problem. Find the information you need.

Decide What to Do and Use Your Ideas

You can use the strategy *Make a Graph* to compare the number of blueberry bagels sold to all the bagels sold.

Step 1 Use the information to start a pictograph.
Each circle stands for ________ bagels.

Step 2 Complete the pictograph.
There are ________ total circles in the pictograph.
There are ________ circles for blueberry bagels.

Is Peggy correct? ________ Explain.

____

Ask Yourself

How many circles should I draw for plain bagels and raisin bagels?

Review Your Work

Look back at the data in the problem to make sure you drew the correct number of circles.

Tell Explain how you could use the numbers in the problem and in the pictograph to answer the question.

____

____

Apply Your Skills

Solve the problems.

2. The pictograph shows how many slices of cheese Len's family used for lunch. They used twice as many slices of turkey as cheese. They used 12 slices of chicken. Did Len's family use more turkey, cheese, or chicken slices?

Sandwich Slices Len's Family Used

Cheese	□ □ □ □
Turkey	
Chicken	

Key: Each □ stands for 2 slices.

Ask Yourself

How many slices of cheese did Len's family use?

Hint Use the key to complete the pictograph.

Answer ______________________________

Identify What words in the problem help you know how many squares to draw for turkey slices?

____

____

3. Lola asked students about their favorite pizza toppings. She found that 5 students like pineapple best. She also found that 4 more students like sausage than like hamburger. She started a bar graph to show the data. How many students in all did Lola ask?

Favorite Pizza Toppings

Toppings: Hamburger, Pineapple, Sausage

0, 5, 10, 15

Number of Students

Ask Yourself

How many students like hamburger best?

Hint Add to find how many students in all Lola asked.

Answer ______________________________

Develop What is another question you can ask using data from this graph?

____

____

Ask Yourself

How can I find the number of peppers, tomatoes, and carrots Emma picks?

Hint Complete the bar graph. Then compare the heights of the bars.

4 Emma picks 4 squash. She picks half as many peppers as squash. She picks 2 times as many tomatoes as squash. She picks 3 times as many carrots as peppers. Which vegetable does Emma pick the most of?

Vegetables from Emma's Garden

Number of Vegetables

10
9
8
7
6
5
4
3
2
1
0

Squash Peppers Tomatoes Carrots

Vegetable

Answer ______________________________

Interpret How is this problem like the Learn About It problem?

____

____

Hint Use the details for Mrs. North's class to help you make a bar graph for Mr. South's class.

Ask Yourself

What operations can I use to find the number of students in Mr. South's class who like each yogurt flavor?

5 Students voted for their favorite yogurt flavor. In Mrs. North's class, 7 students chose vanilla, 4 chose strawberry, and 9 chose chocolate. In Mr. South's class, 3 fewer chose vanilla, 3 more chose strawberry, and 5 fewer chose chocolate. Which flavor was most popular in Mr. South's class?

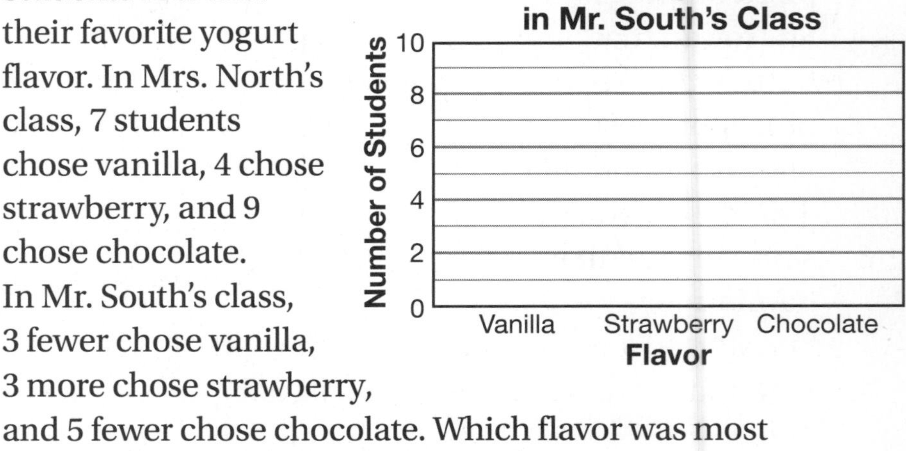

Answer ______________________________

Compare If you made a bar graph for Mrs. North's class, which yogurt flavor would have the tallest bar? The shortest bar?

____

____

On Your Own

Solve the problems. Show your work.

6. Drew asked some students to vote for their favorite sandwich. Ten students voted for peanut butter and jelly. Three fewer students voted for turkey. Only 2 students voted for tuna fish. Seven more students voted for ham and cheese than voted for tuna fish. Which sandwich was the most popular?

Answer ______________________________

Determine Drew forgot to include the number of students who voted for grilled cheese. Eight students voted for grilled cheese. Does this change your answer? Explain.

____

____

7. Ms. Keys wants to know which lunch students like best. Five students who agree on a favorite lunch write their names on 1 paper plate. So 1 plate stands for 5 students. Ms. Keyes uses the plates for a pictograph. There are 6 plates for spaghetti. There are half as many plates for salad as there are for spaghetti. There are 3 times as many plates for pizza as there are for salad. How many students like pizza best?

Answer ______________________________

Sequence What steps did you use to solve the problem?

____

____

Create

Look back at the problems in this lesson. Choose your favorite. Change at least two details in the problem to make your own problem. Solve your new problem.

Lesson 12

Strategy Focus

Make an Organized List

MATH FOCUS: Probability

Learn About It

Read the Problem

Jordan is playing a game with a bag of colored cubes. At the start of the game, the bag has 3 red, 3 orange, 3 green, and 3 blue cubes in it. He takes cubes out of the bag one at a time without looking. He does not put them back. So far, he has taken 2 red, 1 green, and 2 blue cubes out of the bag. What is the probability that the next cube he takes out will be green?

Reread Look for key details. Find the question you need to answer.

- What is this problem about?

____

- What cubes has Jordan taken out of the bag?

____

- What do you need to find out?

____

Mark the Text

Search for Information

Read the problem again. Circle the numbers that you need to use. Underline the question you need to answer.

Record Write what you know about the cubes.

At the start of the game, there are ________ red, ________ orange, ________ green, and ________ blue cubes in the bag.

A lot is happening here. You need a way to organize what you know.

Decide What to Do

You know the number and color of cubes that were in the bag. You know which cubes Jordan has taken out so far.

The probability of an event is the chance that the event will happen. The chance that a coin will be heads when you flip it is 1 out of 2.

Ask How can I find the probability that the next cube will be green?

- I can use the strategy *Make an Organized List* to show all the cubes at the start of the game.
- Next, I can cross out what Jordan has taken out already.
- Then I can look at my new list to count how many cubes are left and how many of them are green.

Use Your Ideas

Step 1 List all the cubes that are in the bag at the start. Use the first letter of each color to make your list.

RRR OOO ________ ________

Step 2 Cross out the letters in your list that stand for the cubes that Jordan has taken out already.

Step 3 The letters that are still on the list show all the possible outcomes for the next cube. Circle the outcomes that are green.

How many letters are still on the list? ________

How many letters stand for green? ________

So the probability that the next cube will be green is ________ out of ________ .

Review Your Work

Look back at your list. Make sure you have crossed out all the cubes that Jordan has taken out.

Explain Ivan thinks the probability is 3 out of 12. He said, "There are 12 cubes, and 3 are green." Explain his error.

____

____

Try It

Solve the problem.

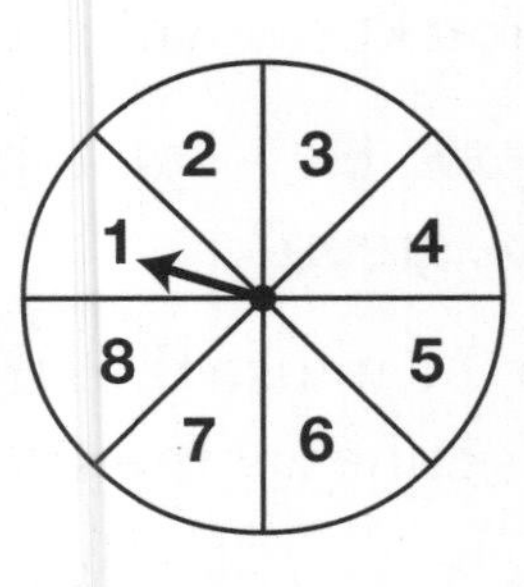

① Zara plays a game with a spinner numbered 1 through 8. If she spins a 3, she gets another turn. If she spins a 7, the other players have to go back to the start. She spins and says, "I hope it's a 3 or a 7!" What is the probability that Zara will get a 3 or a 7?

Read the Problem and Search for Information

Visualize what is happening. Mark key details in the problem.

Decide What to Do and Use Your Ideas

You can use the strategy *Make an Organized List* to keep track of all the outcomes.

Step 1 Zara spins the spinner. List all the possible outcomes.

____

Step 2 Circle the outcomes on your list that Zara wants to get.

Ask Yourself

How many of the possible outcomes are a 3 or a 7?

The probability that Zara gets a 3 or a 7 on her spin is ________ out of ________ .

Review Your Work

Did you count all the possible outcomes? Did you count all the outcomes Zara wants to happen?

Interpret Look at the other possibilities. Is Zara more likely to get one of the other numbers that are *not* 3 or 7? Explain.

____

____

Apply Your Skills

Solve the problems.

② Fred is playing a game with a number cube. If he tosses a 3, he gets 10 points. All the other tosses give him 1 point. What is the probability that Fred will get 10 points on one toss?

Hint A number cube has 6 faces numbered 1 through 6.

Make a list of all the possible outcomes.

____

Ask Yourself

Which of the outcomes should I count to find the probability?

Circle the outcomes that Fred wants.

Answer ______________________________

Contrast What would you do differently to find the probability that Fred does *not* get a 3?

____

____

③ Rick and a friend are playing a game with a spinner. The spinner has 5 equal sections numbered 1 through 5. The player who spins an even number gets to pick a bonus card. What is the probability that Rick will spin an even number?

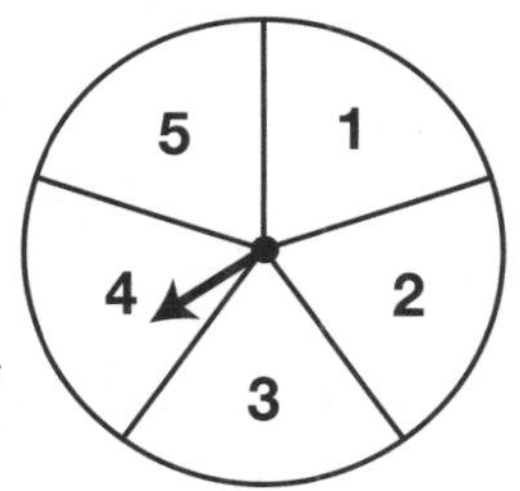

Ask Yourself

How many possible outcomes are there?

List all the possible outcomes.

____

Circle the outcomes the problem asks about.

Hint Circle the even numbers.

Answer ______________________________

Predict What is the probability of getting an odd number? Explain how you know.

____

____

Ask Yourself

How many cards have a circle?

4 Miko is playing a memory game. The cards she has are shown at the right. The 10 cards are mixed up and placed face down so that she cannot see them. Miko flips a card over. What is the probability that it is a circle?

Hint You can use the first letter in the words Square, Circle, Diamond, Heart, and Triangle to write your list of outcomes.

Miko picks a card. List all the possible outcomes.

____

Answer _______________

State How did you decide how many possible outcomes to put on your list?

____

____

5 Amy is playing a board game. She tosses a 1–6 number cube. Then she doubles the number she tossed. She can move that many spaces. What is the probability that Amy gets to move 4 or more spaces on her next turn?

Hint If Amy tosses a 1, she gets to move 2 spaces. If she tosses a 2, she gets to move 4 spaces.

Ask Yourself

What outcomes allow Amy to move 4 or more spaces?

Amy tosses the number cube. List the numbers she could get.

____

Circle the outcomes that allow Amy to move 4 or more spaces.

Answer _______________

Determine What is the probability that Amy gets to move exactly 5 spaces? Explain how you know.

____

____

On Your Own

Solve the problems. Show your work.

6. Mia is playing a word game. She puts the cards shown below in a basket and mixes them up. Without looking, she picks a card from the basket. What is the probability that Mia will pick the letter *B*?

Answer ____________________

Identify Why is the probability of picking the letter *E* the same as picking the letter *B*? Explain.

____

____

7. Yuri and a friend are playing a game with 4 sets of cards. Each set has cards numbered 1 through 5. It is Yuri's turn. He picks one card without looking. The number on the card tells how many spaces to move. What is the probability that Yuri gets to move 4 or more spaces on his turn?

Answer ____________________

Discuss Why is the strategy *Make an Organized List* helpful in solving this problem?

____

____

Create

Choose a problem in this lesson. Change two things about the problem. Write and solve your new problem.

UNIT 3 Review What You Learned

In this unit, you worked with three problem-solving strategies. You can often use more than one strategy to solve a problem. So if a strategy does not seem to be working, try a different one.

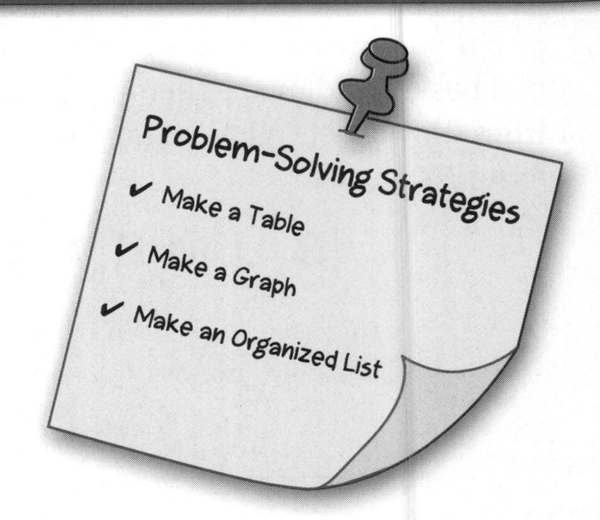

Solve each problem. Show your work. Record the strategy you use.

1. Ms. Thomas had a yard sale. She charged $3 for CDs, $5 for DVDs, and $1 for books. She made a tally chart of what she sold. How much money should Ms. Thomas have from the sale?

CDs	~~||||~~ |||
DVDs	~~||||~~ ||
Books	~~||||~~ ||||

Answer ______________________________

Strategy ______________________________

2. Matt is playing a board game. He tosses a 1–6 number cube. He will send another player back to the start if he rolls a 3 or a 6. Find the probability that Matt will send a player back to the start.

Answer ______________________________

Strategy ______________________________

3. Oscar has a bird feeder. He stays outside for 1 hour each day to count the number of birds that come to the feeder. Here are his counts: 15, 20, 14, 23, 18, 14, 15, 14, 21, 16, 14, 20, 17, 14, 20, 14, 15, 21, 14, 18, 20. What is the middle number for Oscar's counts?

Answer ______________________________

Strategy ______________________________

4. The students in Ms. Jan's class and Mr. Kin's class voted for their favorite color. The graph shows the results for Ms. Jan's class. Mr. Kin's class had:

- 2 votes more for blue
- 4 votes fewer for red
- the same number of votes for yellow

Which color is the most popular in Mr. Kin's class?

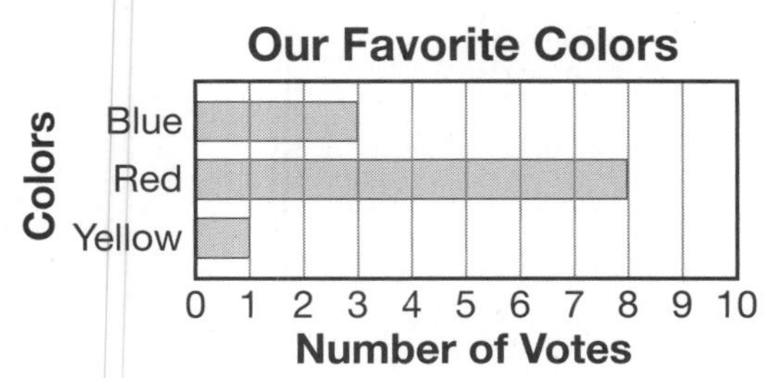

Answer ______________________________

Strategy ______________________________

5. Anna asked 16 people how many pets they have. Six people had 1 dog. Four people had 2 cats. Three people had 1 fish. Three people had 2 birds. How many pets were there in all?

Answer ______________________________

Strategy ______________________________

Explain how you found your answer.

Solve each problem. Show your work. Record the strategy you use.

6. Tomas and Betty are playing a board game with a spinner. The spinner is divided into 12 equal sections, numbered 1 through 12. A player loses a turn if the spinner lands on 3, 8, or 10. Find the probability that Tomas loses a turn when he spins the spinner.

Answer ______________________________

Strategy ______________________________

7. Cows, pigs, and horses have 4 legs. Ducks have 2 legs. Judy went to a farm and saw 1 cow, 3 pigs, 2 horses, and 5 ducks. She counted each animal's legs. How many legs did Judy count in all?

Answer ______________________________

Strategy ______________________________

8. Ana asked her friends how many hours they spend playing outside each week. Here is Ana's data:

7, 14, 16, 13, 1, 14, 4, 1,
10, 6, 13, 5, 17, 4, 4, 9, 10

How many of Ana's friends play outside 7 or more hours each week?

Answer ______________________________

Strategy ______________________________

Explain how you could solve this problem using a different strategy.

9. Donna asked people what kind of pie they liked best. She made the pictograph below to show her results. How many people did Donna ask?

Our Favorite Pies

Apple	🥧🥧🥧🥧🥧
Cherry	🥧🥧
Blueberry	🥧
Pecan	🥧🥧🥧
Other	🥧

Key: Each 🥧 = 5 people.

Answer ______________________

Strategy ______________________

10. Milo has 3 red, 1 green, 4 blue, and 2 yellow marbles in a bag. He takes marbles out of the bag one at a time without looking. He does not put them back. Milo has taken out 1 red marble. What is the probability that the next marble he takes out will be blue?

Answer ______________________

Strategy ______________________

Write About It

Look back at Problem 10. Describe how the strategy you chose helped you to solve the problem.

____

____

Work Together: Make a Graph

Have you ever had a staring contest? Some students recorded how long they could hold their eyes open. Can your team stare longer?

Time (seconds)
42, 46, 5, 31, 10, 12, 28, 50, 23, 34, 24

Plan Use a watch or clock to measure time. Have one person stare. Another person watches for the first blink. A third person keeps the time. Then record the result. Make sure everyone gets a turn at each job.

Create Make a graph in color to show the data you found.

Present Show and explain your graph to the class.

Decide How can you compare your data with others?

UNIT 4 Problem Solving Using Measurement

Unit Theme:
Investigating

What do you investigate? You might want to know what bats do at night. Or maybe you want to find out what size tank you will need for the fish you will buy. In this unit, you will learn how to use math when you are investigating.

Math to Know

In this unit, you will use these math skills:

- Use time and temperature
- Find perimeters and volumes
- Use units of length, weight, and capacity

Problem-Solving Strategies

- Make a Table
- Solve a Simpler Problem
- Guess and Check
- Act It Out

Link to the Theme

Finish the story. Include some of the facts from the table at the right.

Jackie is on a class trip to the aquarium. She makes a table with information about some of the animals there. Jackie shares what she learns with a partner.

Aquarium Animals	Maximum Weight (pounds)
Penguin	9
Sea Lion	300
Octopus	150

Use Math Language

Review Vocabulary

Here are some math words for this unit. Knowing what these words mean will help you understand the problems.

centimeter	inch	ounce	pound
foot	length	perimeter	volume

Vocabulary Activity **Root Words**

Root words can give clues to the meaning of some math terms. Use only two words from the list to complete the following sentences.

1. The root word *meter* means measure. A ________________ is a unit of measure.
2. One ________________ is equal to one hundredth of a meter.
3. The measure of the distance around a figure is called the ________________ .
4. I measured the ________________ of the garden to see how much fence to buy.

Graphic Organizer **Word Diagram**

Complete the graphic organizer.

- Write a definition of the word *length*.
- Draw a diagram to show what *length* means.
- Write a vocabulary term that is an example of a unit of length.
- Write an example of a unit that is *not* a unit of length.

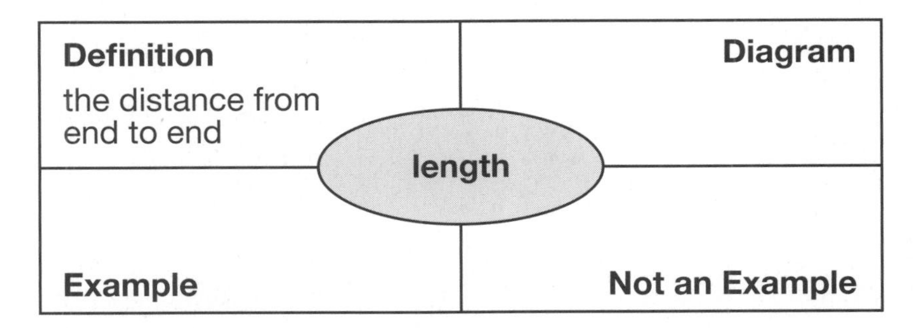

Lesson 13

Strategy Focus

Make a Table

MATH FOCUS: Time and Temperature

Learn About It

Read the Problem

Josh wants to learn more about weather. He starts a weather chart so he can write the high temperature each day. The high temperature was 91°F on Monday. The high temperature went up 2°F each day. What was the high temperature on Friday?

Reread Look back at the problem. Think of these questions as you read.

- What is the problem about?

____

- What do you need to find?

____

Mark the Text

Search for Information

Read the problem again.

Record Find the numbers that will help you solve the problem.

The high temperature on Monday was ________ °F.

The high temperature goes up ________ °F every day.

You can use this information to solve the problem.

Decide What to Do

You know the temperature on Monday. You know the temperature goes up by 2°F each day.

Ask How can I find the high temperature on Friday?

- I can use the strategy *Make a Table* to keep track of the high temperatures each day.
- I will use the high temperature on Monday to find the high temperature on Tuesday. Then I can find the high temperatures for the rest of the days.

Use Your Ideas

Step 1 Make a table showing the days and high temperatures. Start by writing the days.

Day	Monday	Tuesday	Wednesday		
High Temperature					

The days of the week you need in your table are Monday, Tuesday, Wednesday, Thursday, and Friday.

Step 2 Write Monday's temperature. Add 2°F to find Tuesday's temperature. Keep adding 2°F until you get to Friday.

Day	Monday	Tuesday	Wednesday	Thursday	Friday
High Temperature	91°F	93°F	95°F		

So the high temperature on Friday was ________________ .

Review Your Work

Check that the temperature each day after Monday is 2°F more than the day before.

Describe Suppose the high temperature continues to increase by 2°F each day. How can you use the table to find out when the temperature will reach 105°F?

____

____

Try It

Solve the problem.

① Ms. Thomas wants to learn about what bats do at night. She sets a timer to take a picture every 15 minutes. She starts taking pictures at 11:30 P.M. When is the sixth picture taken?

Read the Problem and Search for Information

Think about what you need to find out. Read the problem again. Circle the words and numbers that will help you.

Decide What to Do and Use Your Ideas

You can use the strategy *Make a Table* to keep track of the number of pictures and the time.

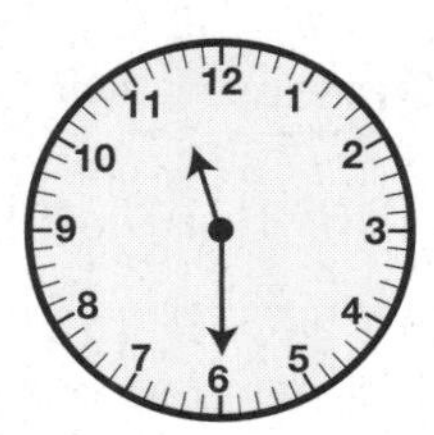

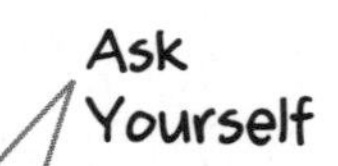

How should I set up the table?

Step 1 Write the picture numbers in the top row.

Step 2 Write the time each picture is taken. Use the clock to help. Find the time that the sixth picture is taken.

Picture	1	2	3			
Time	11:30 P.M.	11:45 P.M.	12:00 A.M.			

So the sixth picture will be taken at ________________.

Review Your Work

Check that you answered the question asked.

Compare How is this problem similar to and different from the Learn About It problem?

____

____

Apply Your Skills

Solve the problems.

② Kim is comparing desert temperatures. She finds the average high and low summer temperatures for three deserts. The table shows her data for Desert 1 and Desert 2. The high in Desert 3 is 40°C. The low is 32°C. Which desert has the greatest difference between high and low summer temperatures?

Hint A table can show the temperatures in each desert.

	Average High Summer Temperature	Average Low Summer Temperature	Difference
Desert 1	50°C	40°C	10°C
Desert 2	45°C	20°C	
Desert 3			

Ask Yourself

How can I find the differences in the temperatures?

Answer ______________________________

Explain How did making a table help you solve the problem?

____

____

③ Five teams of scientists will search the Sahara Desert for ancient fossils. They each go out in desert cars. One team will go out 45 minutes after another team. The first team goes at 11:00 A.M. Mr. Finn is with the fifth team that goes. What time does his team go?

Hint Use a clock to help.

Team	1	2			
Time	11:00 A.M.	11:45 A.M.			

Ask Yourself

If the second team goes at 11:45 A.M., at what time will the third team go?

Answer ______________________________

Clarify Jack thinks Mr. Finn's team goes at 2:45 P.M. What might his error have been? Explain what Jack needs to do.

____

____

Ask Yourself

The cloud is getting closer. Do I need to add or subtract?

Hint You can end your table when you find the answer.

④ José and Eva live in a desert area. They are recording cloud movements. A storm cloud floats toward them. It is 15 miles away. It gets 3 miles closer every 5 minutes. How long will it take for the cloud to reach them?

Time (minutes)	0	5				
Cloud Distance (miles)	15					

Answer ______________________________

Interpret How did you know when you found the answer?

____

____

Hint Noon is 12:00 P.M.

⑤ Dr. Weiss studies temperatures in the desert. He compares the temperatures in the sun and in the shade. It is 100°F in the sun at noon. This temperature rises 3°F every hour until 3:00 P.M. Dr. Weiss finds that the temperature is always 10°F cooler in the shade. What was the temperature in the shade at 2:00 P.M?

Ask Yourself

Do I need to end the table at 3:00 P.M.?

Time	Noon		
Temperature in the Sun	100°F		
Temperature in the Shade	90°F		

Answer ______________________________

Determine Suppose you did *not* include a row for the temperature in the shade. How could you find the answer without that row?

____

____

On Your Own

Solve the problems. Show your work.

6. Mitch and his classmates go to a cactus park. They will observe the different colors, shapes, and sizes of the cactus plants. The class leaves the ranger hut at 9:00 A.M. to follow the trail. Every 40 minutes, they pass a rest station. What time do they pass the third rest station?

Answer ______________________________

Identify If the class goes back to the ranger hut after the third rest station, how could you use a table to find what time they get back to the hut?

____

____

7. Ms. Olu tracks the weather in the desert every day. She reads the thermometer and records the high temperatures. On Monday, it was 94°F. It was 4°F more on Tuesday than on Monday. It was 1°F less on Wednesday than on Tuesday. It was 6°F more on Thursday than on Wednesday. What was the temperature on Thursday?

Answer ______________________________

Analyze How did you know whether to add to or subtract from the temperatures?

____

____

Create

Look back at the problems in the lesson. Write a problem about the temperature in the desert. Your problem should be one that can be solved by making a table. Solve your problem.

Strategy Focus

Solve a Simpler Problem

MATH FOCUS: Length, Weight, and Capacity

Learn About It

Read the Problem

Workers want to put 5 jellyfish into tanks. The jellyfish are different sizes. The tanks are different sizes, too. The largest jellyfish needs the largest tank. The names and sizes of the jellyfish are in the table below. Put the jellyfish in order from smallest to largest.

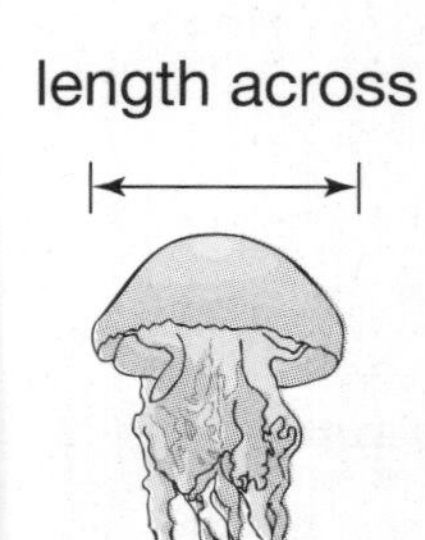

Jellyfish	Average Length Across
Man	10 inches
Moon	15 inches
Noma	79 inches
Spot	18 inches
Win	7 inches

Reread Ask yourself questions as you read.

- What is the problem about?

____

- What are you asked to do?

____

Search for Information

Read the problem again. Circle information you need.

Record Write what will help you solve the problem.

Workers want to put ________ jellyfish into tanks.

The jellyfish are ________________ sizes.

Think about a way to use what you know about the numbers to solve this problem.

Decide What to Do

You need to order the lengths. The lengths are all in inches. There are 5 numbers to compare.

Ask How can I put the jellyfish in order from smallest to largest?

- I can use the strategy *Solve a Simpler Problem.*
- I can find the least and greatest numbers. Then I can compare two numbers at a time until I have ordered all the numbers.

Use Your Ideas

Step 1 Make a list. Look for numbers you can place in your list right away. Two of the numbers are 10 or less. The number__________is much greater than the other numbers.

7, 10, ?, ?, 79

Step 2 Compare the numbers that are left.

15 is less than 18. Write _________ to the right of 15.

7, 10, _________, _________, 79

So the jellyfish in order from least to greatest are

_________, _________, _________, _________, _________.

You do not have to compare the numbers all at once.

Review Your Work

Read the problem again. Did you answer in the order the problem asks for?

Explain How does the strategy *Solve a Simpler Problem* help you order the numbers?

____

____

Try It

Solve the problem.

1. A scientist studies turtles. She wants to know how many pounds of sea sponges a turtle can eat. She learns that 8 turtles can eat about 4,800 pounds of sponges in 6 months. About how many pounds of sponges can 1 turtle eat in 6 months?

Read the Problem and Search for Information

Mark numbers and words that you need to solve the problem.

Decide What to Do and Use Your Ideas

You can use the strategy *Solve a Simpler Problem.*

Ask Yourself

Will the amount that 1 turtle can eat be more than or less than the amount that 8 turtles can eat?

Step 1 You know how many pounds of sea sponges ________ turtles can eat.

You need to find how many pounds ________ turtle can eat.

So you should divide 4,800 pounds by 8.

Step 2 Use simpler numbers to help you divide. Use basic facts and patterns of zeros.

$48 \div 8 =$ ________

$480 \div 8 =$ ________

$4{,}800 \div 8 =$ ________

So 1 turtle can eat about ________ pounds of sea sponges in 6 months.

Review Your Work

Multiply to check your answer.

Identify How does using basic facts and patterns of zeros make this a simpler problem?

____

____

Apply Your Skills

Solve the problems.

2 Nan is studying fish. She measures the lengths of some fish. Which is the third longest fish?

Fish	Length
Fish A	20 centimeters
Fish B	7 centimeters
Fish C	25 centimeters
Fish D	4 centimeters
Fish E	60 centimeters

Ask Yourself
Which are the greatest and the least numbers?

Length in centimeters of the fish in order from longest to shortest: 60 cm, ________, ________, ________, 4 cm

Fish in order from longest to shortest: ________, ________, ________, ________, ________

Hint The third longest fish is the third fish in your list.

Answer ______________________________

Describe How does solving a simpler problem help you list the numbers?

____

____

3 Cho is comparing the amount of water in two tanks. Cho has a fish tank that holds 12 gallons of water. Seaside Park has a fish tank that holds 4,000 times as much water. How much water does the Seaside Park fish tank hold?

Ask Yourself
Suppose the Seaside Park tank held only 4 times as much as Cho's tank. What would I do?

$4 \times 12 = 48$

$40 \times 12 =$ ________

$400 \times 12 =$ ________

$4{,}000 \times 12 =$ ________

Hint Use multiplication and patterns of zero.

Answer ______________________________

Interpret How do the words *times as much* help you choose an operation?

____

____

Ask Yourself

How can I use simpler numbers to compare 9 feet to 40 inches?

Hint Remember that 1 foot is equal to 12 inches.

4. Jamil is writing a report about sharks. He reads about their lengths. Nurse sharks grow to be about 9 feet long. Horn sharks grow to be about 40 inches long. Which shark is longer when it is fully-grown?

1 foot is about ________ inches.

So 9 feet is about ________ inches.

Answer ____

Consider Can you change inches to feet to compare the length of the sharks and still get the same answer? Explain.

____

____

5. Mike has a fish tank. The fish are different colors. The table shows how much they weigh. Is Mike right when he says the red fish is his second lightest fish? Explain.

Fish Color	Weight
White	77 ounces
Brown	103 ounces
Red	56 ounces
Orange	77 ounces
Green	39 ounces

Hint You can compare two weights at a time to list them from lightest to heaviest.

The weight of the fish in ounces in order from lightest to heaviest:

______, ______, ______, ______, ______

Fish in order from lightest to heaviest:

____________, ____________, ____________,

____________, ____________

Ask Yourself

Does the second lightest fish weigh more than or less than the lightest fish?

Answer ____

Conclude Why would it be easy for Mike to tell which fish weighs the most?

____

____

On Your Own

Solve the problems. Show your work.

⑥ Some scientists are taking care of a baby blue whale. It drinks about 400 liters of milk each day. Is 1,800 liters of milk enough to feed a baby blue whale for 6 days? Why or why not?

Answer ______________________________

Determine For how many full days would 1,800 liters of milk be enough to feed a baby blue whale? Explain.

____

____

⑦ Mrs. Tran looked at the sign at the pet store. Then she bought the second-longest fish for her saltwater tank. How much did Mrs. Tran pay for her new fish?

Fish	Length	Price
Eel	24 inches	$17
Cleaner Shrimp	2 inches	$28
Seahorse	11 inches	$48
Cat Shark	48 inches	$50
Squirrel Fish	9 inches	$19

Answer ______________________________

Compare Suppose Mrs. Tran bought the fish that cost the least instead of the second longest. What would be different about how you solve this problem?

____

____

Create

Make up a problem about the weights of baby sea turtles. Your problem should be one that can be solved using the strategy *Solve a Simpler Problem*. Write and solve your problem.

Strategy Focus

Guess and Check

MATH FOCUS: Perimeter

Learn About It

Read the Problem

Mr. Lo wants to make a dollhouse for his daughter. He is making the plans. The floor of the kitchen will be shaped like a rectangle. It will have a perimeter of 36 inches. The long wall will be 12 inches long. What will be the length of the kitchen's short wall?

Plan for a Dollhouse Kitchen

Reread Ask yourself questions while you read.

- What is the problem about?

____

- What kind of information do you know?

____

- What do I need to find?

____

Mark the Text

Search for Information

Read the problem again. Find the important numbers.

Record What numbers are used in the plan?

The perimeter of the kitchen floor will be ________ inches.

A long wall of the kitchen will be ________ inches long.

Think about how this information can help you solve the problem.

Decide What to Do

The problem tells you what the perimeter of the kitchen floor will be. It also tells you the length of its long wall.

The perimeter is the distance around a shape.

Ask How can I find the length of the kitchen's short wall?

- I can use the strategy *Guess and Check.*
- I can guess a number for the length of a short wall. Then I can check if the perimeter will be 36 inches. If not, I can make another guess.

Use Your Ideas

Step 1 Try 10 inches for the length of the short wall.

	Short Wall (inches)	Long Wall (inches)	Perimeter (inches)
Guess 1	10	12	10 + 10 + 12 + 12 = 44

Step 2 Try a length that is less than 10 inches.

	Short Wall (inches)	Long Wall (inches)	Perimeter (inches)
Guess 2	5	12	5 + 5 + 12 + 12 = ☐

Step 3 Try a length that is a little longer than 5 inches.

	Short Wall (inches)	Long Wall (inches)	Perimeter (inches)
Guess 3	6	12	6 + 6 + ☐ + ☐ = ☐

So the short wall of the kitchen will be ________ inches long.

Review Your Work

Check that you answered the question the problem asked.

Describe How can you use the perimeter from one guess to make your next guess?

____

____

Try It

Solve the problem.

1. Mary has a garden with 5 sides. She drew this picture of her garden. The perimeter is 30 feet. Side *A* is 2 feet shorter than Side *B*. What are the lengths of Sides *A* and *B*?

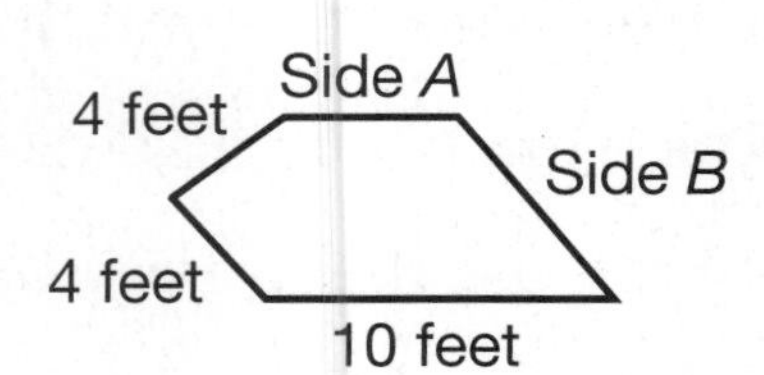

Read the Problem and Search for Information

Identify what you know and what you need to find out.

Decide What to Do and Use Your Ideas

You can use the strategy *Guess and Check.* Complete the table as you work through the steps.

Ask Yourself

How can I use what I know to find the lengths I do not know?

Step 1 Add the lengths of the 3 sides you know.

10 feet + 4 feet + ________ feet = ________ feet

Step 2 Try 3 for the length of Side *A*. Add 2 feet for Side *B*. Add all the lengths. Compare your result to 30.

	3 Other Sides (feet)	Side *A* (feet)	Side *B* (feet)	Perimeter (feet)
Guess 1	18	3	5	18 + ☐ + 5 = 26
Guess 2	18	6		18 + ☐ + ☐ = ☐
Guess 3	18	5		

Step 3 Side *A* must be longer than 3 feet. Try 6 feet.

Step 4 Side *A* must be shorter than 6 feet. Try 5 feet.

So Side *A* is ________ feet long. Side *B* is ________ feet long.

Review Your Work

Check that your answer makes sense.

Generalize How does a table help you organize your guesses?

____

____

Apply Your Skills

Solve the problems.

② Sam is building a fort shaped like a rectangle. He puts rocks around the edges. He puts the same number of rocks along the front and back. He puts 4 fewer rocks along each of the other two sides. Sam uses a total of 32 rocks. How many rocks does he use for the front?

	Rocks Along Front	Rocks Along One Side	Total Number of Rocks Front + Back + Side + Side
Guess 1	5	1	5 + 5 + 1 + 1 = 12
Guess 2	15	11	15 + 15 + 11 + 11 = ☐
Guess 3	10		10 + 10 + ☐ + ☐ = ☐

Answer ______________________

Explain Suppose your guess for the rocks along the front and back was 8. How could you use the answer from that guess to help you make your next guess?

____

____

Ask Yourself

What is a good first guess?

Hint There are 4 fewer rocks on the two other sides. So there must be at least 5 rocks on the front and back.

③ Leo is making a kite. The two short sides are the same length. The two long sides are the same length. The short sides are 3 inches shorter than the long sides. A border around the sides uses 26 inches of ribbon. How long are the sides of his kite?

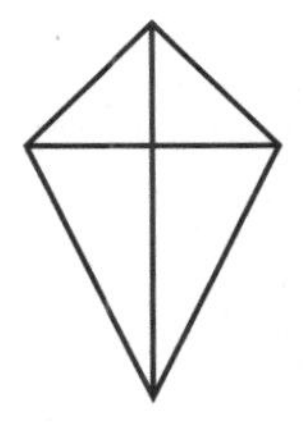

	Long Sides (inches)	Short Sides (inches)	Total Border (inches)
Guess 1	10	7	
Guess 2	5		
Guess 3	8		

Answer ______________________

Identify Tell why 50 inches is *not* a good first guess.

____

____

Ask Yourself

How does the length of the border help me find the side lengths?

Hint Choose a length for the long side of the kite.

Ask Yourself

Draw a picture to help you understand the problem.

Hint Are the front and the back the short or the long sides of the room?

④ Ali is putting a border around her room. The floor of her room is shaped like a rectangle. The perimeter of the floor is 48 feet. The front and back are each 2 feet shorter than the other sides. Ali finished putting the border on the front, the back, and one other side. How many feet of border does she need for the last side?

	Long Sides (feet)	Front and Back (feet)	Perimeter (feet)
Guess 1	10		
Guess 2	15		
Guess 3			

Answer ________________________

Relate Could you make guesses for the front and back instead of the long sides? Explain.

____

____

Ask Yourself

How do I find the perimeter of this shape?

Hint Do not forget the fifth side when finding the perimeter.

⑤ Laura's art project is due next week. The design for her project looks like this. Sides *A* and *B* have the same length. Sides *C* and *D* have the same length. Side *C* is twice as long as Side *A*. The distance around the design is 39 inches. How long are the sides of her design?

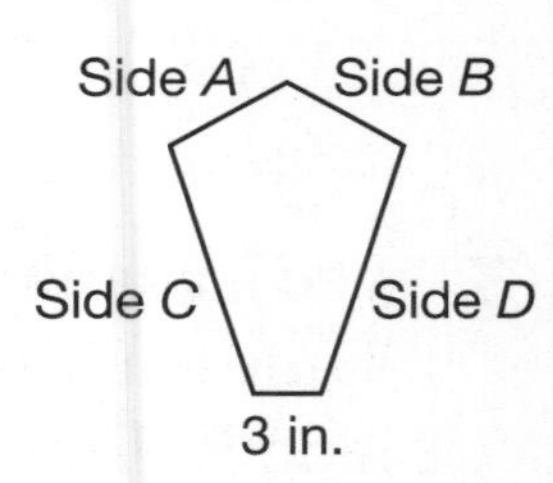

	Side *A* and *B* (inches)	Side *C* and *D* (inches)	Perimeter (inches)
Guess 1	10		
Guess 2			
Guess 3			

Answer ________________________

Compare How is this problem like Problem 1?

____

____

On Your Own

Solve the problems. Show your work.

6. Hector is looking at plants growing around the outside of his house. The shape of the bottom of his house is a rectangle. He walks 30 feet along the short side. It is 180 feet around the whole house. How many feet does Hector walk along the long side?

Answer ____

Justify What number did you use for a first guess? Why?

____

____

7. Ryan's dad is painting a ceiling that looks like this. He has this information about the sides.

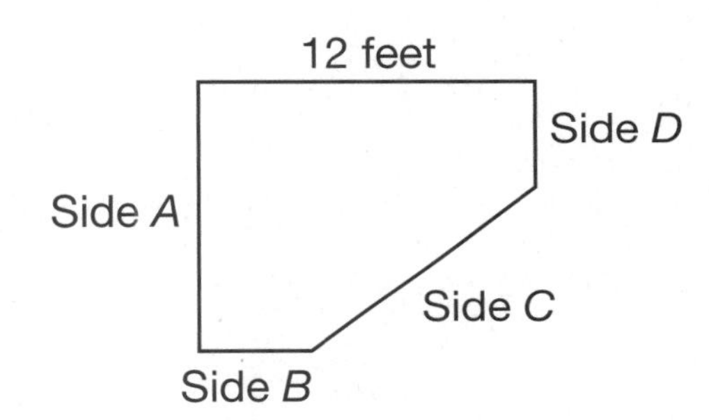

Sides *A* and *C* have the same length.
Sides *B* and *D* have the same length.
Side *B* is 6 feet shorter than Side *A*.

He puts up 40 feet of tape around the edge of the ceiling. The tape is to keep the paint off the walls. How long are Sides *C* and *D*?

Answer ____

Formulate What is another question you could ask about this problem?

____

____

Create

Look back at the problems in the lesson. Write a problem about placing a border around something. Your problem should be one that can be solved using *Guess and Check*. Then solve your problem.

Lesson 16

Strategy Focus

Act It Out

MATH FOCUS: Volume

Learn About It

Read the Problem

Alba grows flowers on vines. She wants to build this wall for the vines to grow on. She has 50 bricks. The bricks are 1 foot long, 1 foot high, and 1 foot wide. How many bricks will she need to build the wall?

$h = 4$ feet

$w = 1$ foot

$l = 10$ feet

Reread Think about these questions as you read.

- What is the problem about?

____

- Where will I get the numbers I need?

____

- What question do I have to answer?

____

Search for Information

Look at the facts in the problem and in the picture.

Record What numbers do you know?

Alba has ________ bricks.

Each brick is ________ foot long, ________ foot high, and ________ foot wide.

The wall will be ________ feet long, ________ feet high, and ________ foot wide.

Think about how you can use these facts to solve the problem.

Decide What to Do

You know the size of Alba's bricks. You also know the size of Alba's wall.

Ask How many bricks will she need to build the wall?

- I can use the strategy *Act It Out* to create a model of Alba's wall.
- I can use centimeter cubes or connecting cubes to stand for the bricks in the wall. Then I can find how many cubes there are in all.
- The number of bricks needed is the volume of the wall.

Use Your Ideas

Step 1 The wall is 10 feet long. So make one row of 10 cubes.

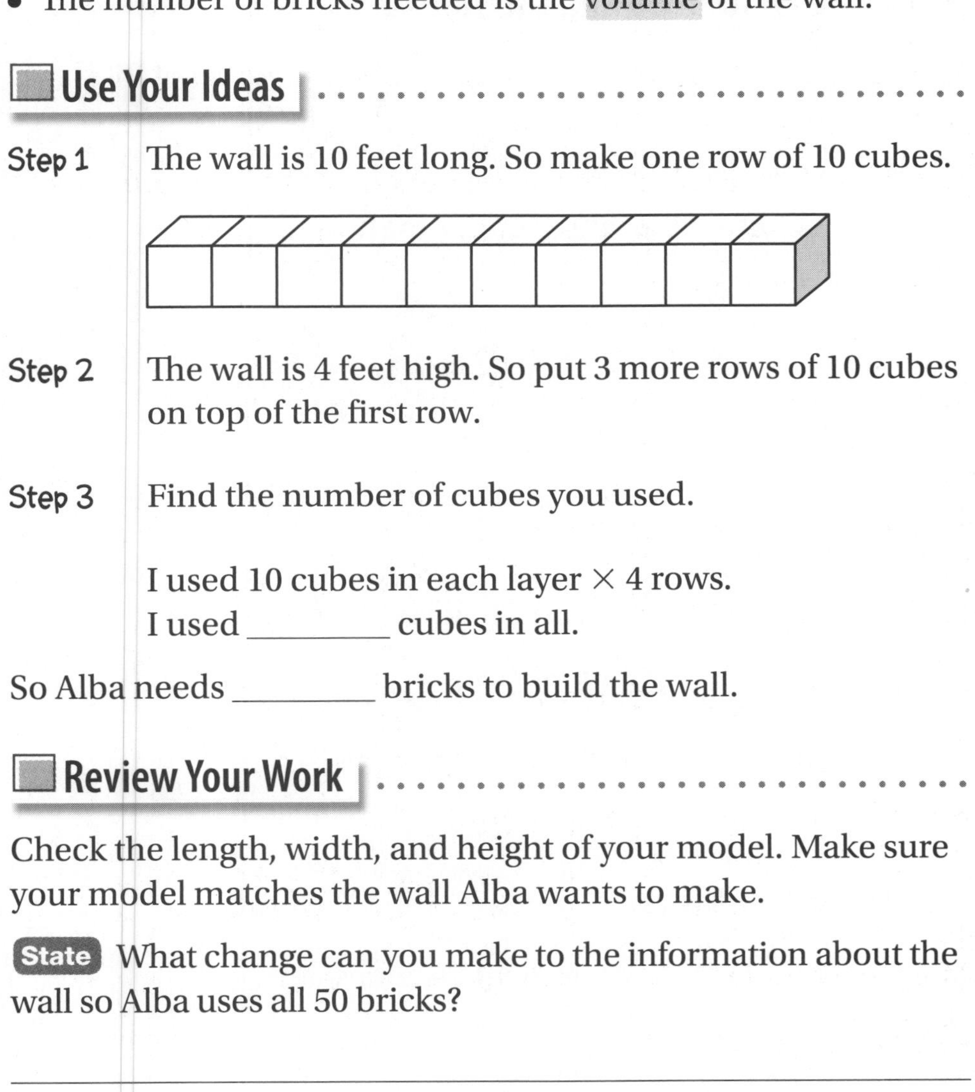

Step 2 The wall is 4 feet high. So put 3 more rows of 10 cubes on top of the first row.

Step 3 Find the number of cubes you used.

I used 10 cubes in each layer × 4 rows.
I used ________ cubes in all.

So Alba needs ________ bricks to build the wall.

Review Your Work

Check the length, width, and height of your model. Make sure your model matches the wall Alba wants to make.

State What change can you make to the information about the wall so Alba uses all 50 bricks?

____

____

Try It

Solve the problem.

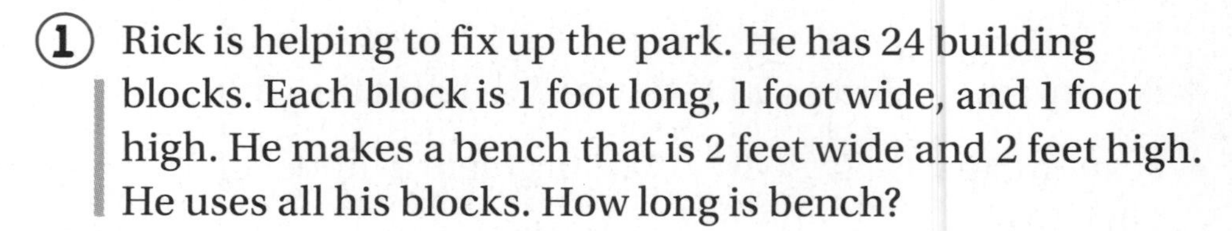

(1) Rick is helping to fix up the park. He has 24 building blocks. Each block is 1 foot long, 1 foot wide, and 1 foot high. He makes a bench that is 2 feet wide and 2 feet high. He uses all his blocks. How long is bench?

Read the Problem and Search for Information

Reread the problem. Mark the numbers you need.

Decide What to Do and Use Your Ideas

You can use the strategy *Act It Out* to create a model of Rick's bench. You can use cubes to stand for the blocks.

Ask Yourself

How wide and how high is the bench?

Step 1 Make a small part of the bench. Make it 1 cube long, 2 cubes wide, and 2 cubes high.

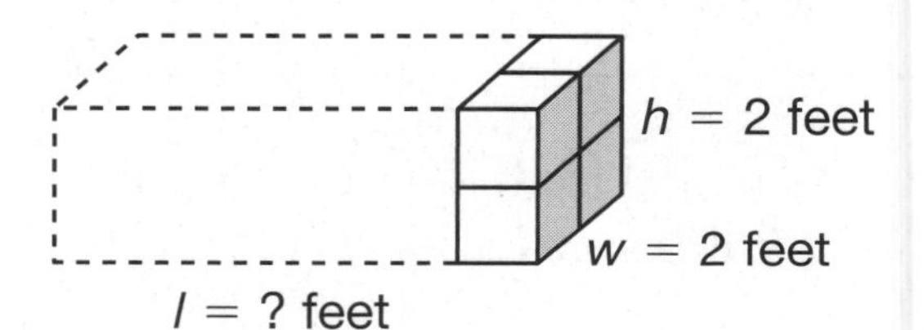

Step 2 Add another small part of the bench.

Step 3 Repeat Step 2 until you have no more cubes.

Step 4 Count how many cubes long your model is.

The model is ________ cubes long.

So the bench is ________ feet long.

Review Your Work

Make sure that you have used all 24 cubes.

Conclude How did using the strategy *Act It Out* help you solve the problem?

____

____

Apply Your Skills

Solve the problems.

② Bart is putting away 45 number cubes after class. He wants to fill a box with these cubes. The box fits 3 cubes along the length. It fits 2 cubes along the width. Five layers of cubes fit in the box. How many number cubes can Bart put in the box?

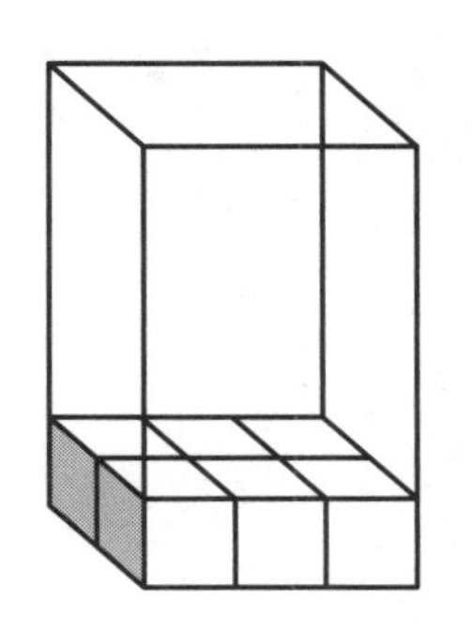

Hint Use cubes to show the first layer in the box.

Each layer has ________ cubes.

________ layers are needed to fill the box.

Ask Yourself

How high is the box?

Answer ______________________________

Identify Roman thinks the answer is 45. What is his error? Explain what he needs to do.

____

____

③ Sarah is going to make a plate out of clay. The clay comes in cubes 1 inch long, 1 inch wide, and 1 inch high. A box holds 48 of the clay cubes. The box is 4 inches long and 3 inches wide. How high is it?

Ask Yourself

How many layers of clay cubes will the box hold?

A box holds ________ clay cubes.

Each layer has ________ clay cubes.

Answer ______________________________

Hint You can divide the total number of cubes by the number of cubes in each layer to check your answer.

Explain How did you find the number of clay cubes there are in each layer?

____

____

4 Lulu buys notepads at a gift store. They are shaped like cubes. They are sold in boxes. Each box is 5 pads long, 4 pads wide, and 2 pads high. How many notepads are in a box?

Hint Use cubes to act out the problem.

How many notepads are in each layer? ________

How many layers are in a box? ________

Ask Yourself

What can I do to see how they fit in a box?

Answer ________________________________

Rearrange Suppose each box is 4 pads long, 2 pads wide, and 5 pads high. Would your answer change? Explain how you know.

____

____

5 Art is moving. He is packing his things in boxes like the one shown. The boxes will go in an empty corner of a room until moving day. The space in the corner is 3 yards long, 3 yards wide, and 3 yards high. How many boxes will fit in the corner?

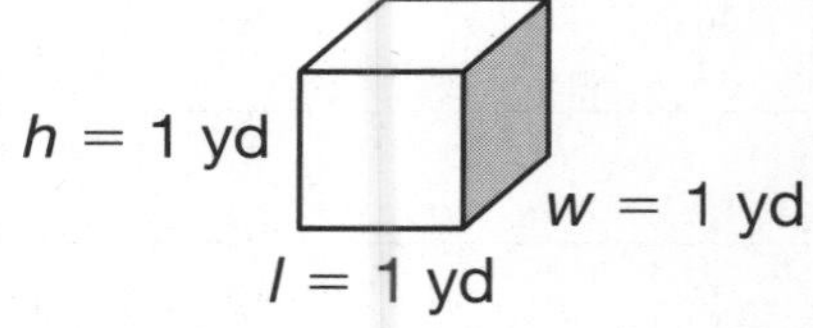

Hint Find how many boxes can sit directly on the floor in the corner.

How many boxes will be in each layer? ________

How many layers of boxes will be in the corner? ________

Ask Yourself

What does the height of the corner space tell me about how many layers of boxes there will be?

Answer ________________________________

Compare How is this problem like Problem 2?

____

____

On Your Own

Solve the problems. Show your work.

6 Meg is 3 years old. She has a new box of wood blocks. The picture shows the box. It holds 64 blocks. How many blocks high is the box?

h = ? blocks

64 wood blocks

w = 2 blocks

l = 8 blocks

Answer ______________________________

Judge What information in the problem was *not* needed?

____

____

7 Dan's fish tank is 4 feet long, 2 feet wide, and 3 feet high. He wants to find the volume of the tank. He finds cubes that are 1 foot long, 1 foot wide, and 1 foot high. How many cubes will Dan need to fill the fish tank?

Answer ______________________________

Modify Ari wants to build a tank with the same volume. It will be a different size. How many feet long, wide, and high can his new tank be?

____

____

Create

Write a problem that can be solved by using centimeter or connecting cubes to act it out. Make your problem about using bricks to build a wall. Then solve your problem.

Review What You Learned

In this unit, you worked with four problem-solving strategies. You can often use more than one strategy to solve a problem. So if a strategy does not seem to be working, try a different one.

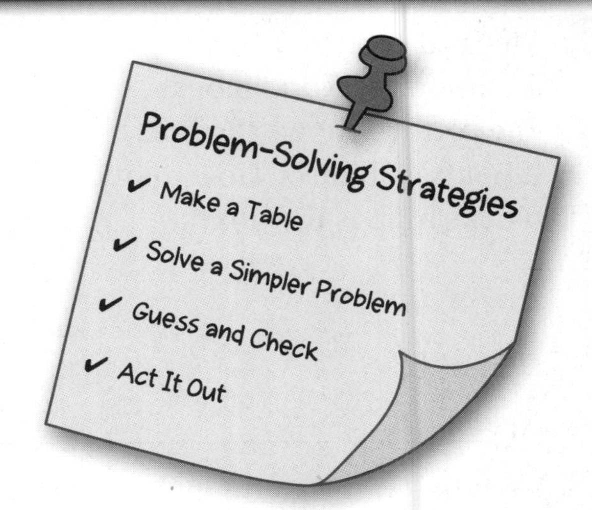

Solve each problem. Show your work. Record the strategy you use.

1. Rona is making bread. She wants to find out when it will be done. It will be ready when the temperature of the bread is at least 210°F. The temperature of the bread is 70°F when Rona puts it in the oven at 10:00 A.M. The temperature of the bread goes up 35°F every 10 minutes. When will the bread be ready?

Answer ____________________

Strategy ____________________

2. Suppose a mother rhino weighs 3,600 pounds. It weighs 40 times as much as its baby. How much does the baby rhino weigh?

Answer ____________________

Strategy ____________________

3. Alex has 74 feet of wire fence. He has just enough to go all around the edge of his garden once. His garden is shaped like a rectangle. It is 13 feet longer than it is wide. How long and how wide is Alex's garden?

Answer ______________________________

Strategy ______________________________

4. Mrs. Kim teaches math. She has these two boxes to store centimeter cubes. She wants to find out which box holds more cubes. Which box will hold more cubes? How many more?

A: 4 cm, 2 cm, 3 cm

B: 5 cm, 2 cm, 2 cm

Answer ______________________________

Strategy ______________________________

5. The pet store has five snakes. The table shows how long each snake is. Which snake is the second shortest snake?

Snake	Length
Corn Snake	77 centimeters
Garter Snake	54 centimeters
Rat Snake	92 centimeters
King Snake	36 centimeters
Ribbon Snake	48 centimeters

Answer ______________________________

Strategy ______________________________

Explain how you found your answer.

Solve each problem. Show your work. Record the strategy you use.

6. Jill saw that the temperature at 7 P.M. was 81°F. It cooled by 3°F each hour until 11 P.M. Then it cooled 1°F each hour until 5 A.M. What was the temperature when she woke up at 5 A.M.?

Answer ______________________

Strategy ______________________

7. The floor of a play area is the shape of a rectangle. The floor is 8 feet longer than it is wide. The perimeter of the floor is 40 feet. How long is the floor? How wide is it?

Answer ______________________

Strategy ______________________

8. Mara is making a map of her local park. The shape of the park is shown below. Side *B* is the longest side. It is 3 times the length Side *A*. The path along the outside of the park is 680 feet long. How long are Sides *A* and *B*?

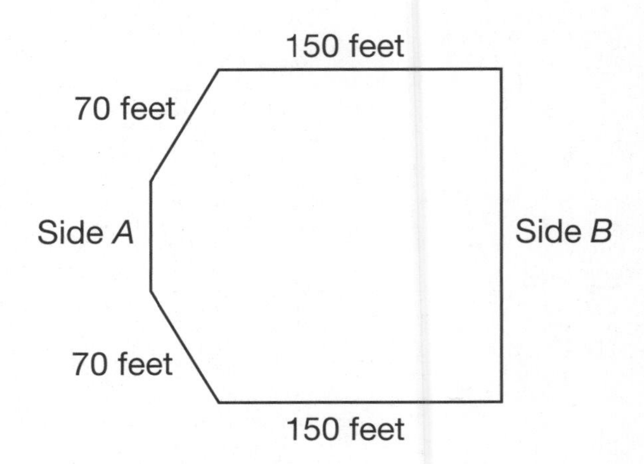

Answer ______________________

Strategy ______________________

Explain how you know the lengths of the sides of the park.

9. Kate's toy box is 2 feet high, 5 feet long, and 3 feet wide. She is going to fill the toy box with cubes that are 1 foot on each side. How many cubes does she need?

Answer ______________________

Strategy ______________________

10. Rex sees 3 little brown bats flying in the sky. Rex knows that a little brown bat can eat about 600 insects in one hour. About how many insects can 3 little brown bats eat in one hour?

Answer ______________________

Strategy ______________________

Write About It

Look back at Problem 7. Explain your reasoning.

____

____

Work Together: Design a House Floor Plan

Your team needs to design a floor plan for a one-story house. A floor plan shows what the floor of the house looks like when it is looked at from above. The shape of the floor of the house will be a rectangle. Draw a floor plan showing a living room, kitchen, bathroom, and two bedrooms.

Plan Work as a group. Think about where you want the rooms to be.

1. Draw a rectangle for the floor of the house.
2. Show where the walls and doors are. Be sure to include a hallway.

Decide Choose the length and width of the house.

Create Make a large drawing of your plan. Include the perimeter of each room and the perimeter of the whole house.

Present Share your drawing with the class. Explain why you decided on this plan.

UNIT 5 Problem Solving Using Geometry

Unit Theme: Art in Our World

Art comes in many shapes, sizes, and forms. People use shapes to design jewelry, book covers, or T-shirts. You can use shapes to build sand castles or to paint pictures. In this unit, you will see how math is used in all kinds of art.

Math to Know

In this unit, you will use these math skills:

- Identify polygons and their properties
- Recognize slides, flips, and turns
- Identify congruent figures
- Recognize and name solid shapes

Problem-Solving Strategies

- Draw a Picture
- Act It Out
- Guess and Check
- Make an Organized List

Link to the Theme

Finish the story. Tell how Julie describes her necklace. Include some of the words from the list at the right.

Julie wants to make a necklace in art class. She draws her design on a sheet of paper. Julie tells her mom about the design when she gets home.

____

____

____

____

____

____

____

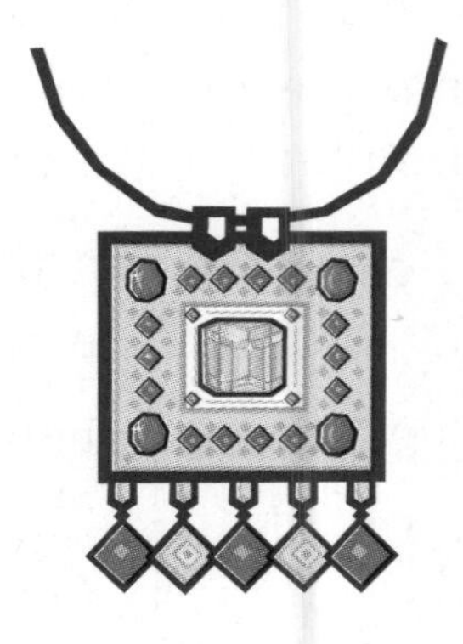

Words to Use

shapes	smaller
square	larger
octagon	

Use Math Language

Review Vocabulary

Here are some math words for this unit. Knowing what these words mean will help you understand the problems.

edge	horizontal	rhombus	triangle
face	polygon	trapezoid	vertical

Vocabulary Activity **Word Pairs**

Math words that are often learned together may be related. Use words from the above list to complete the following sentences.

1. The flat surface of a solid object is called a ________________ .
2. Two faces of a solid figure meet at an ________________ .
3. A ________________ line goes left to right.
4. A ________________ line goes up and down.

Graphic Organizer **Word Web**

Complete the graphic organizer.

- In the large center oval, draw an example of a polygon.
- In each empty oval, write a different vocabulary word that relates to the word *polygon*.

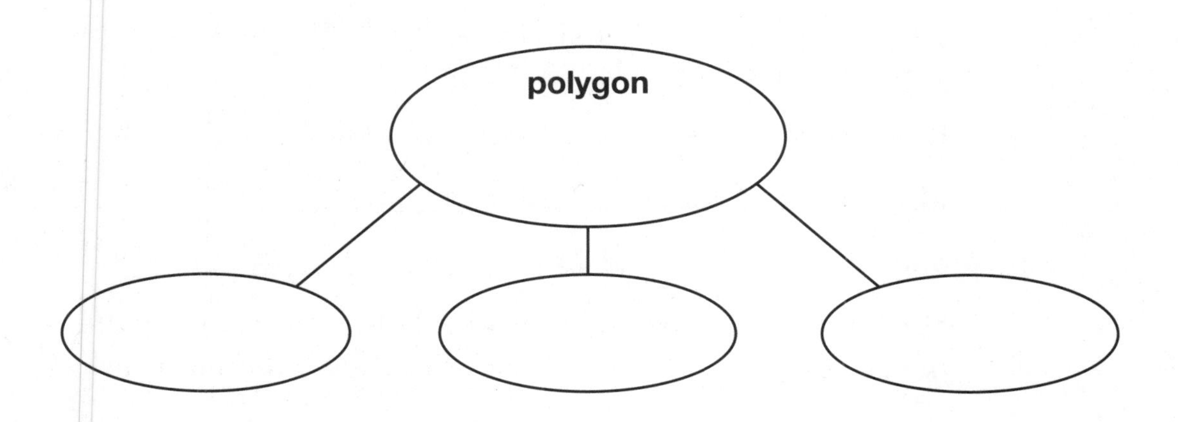

Lesson 17

Strategy Focus

Draw a Picture

MATH FOCUS: Polygons

Learn About It

Read the Problem

Kim wants to make a 10-foot by 8-foot rug. She will use 5 small pieces of rug. One piece is a rectangle. Two pieces are squares. Two pieces are right triangles. How can she make these pieces into a 10-foot by 8-foot rug?

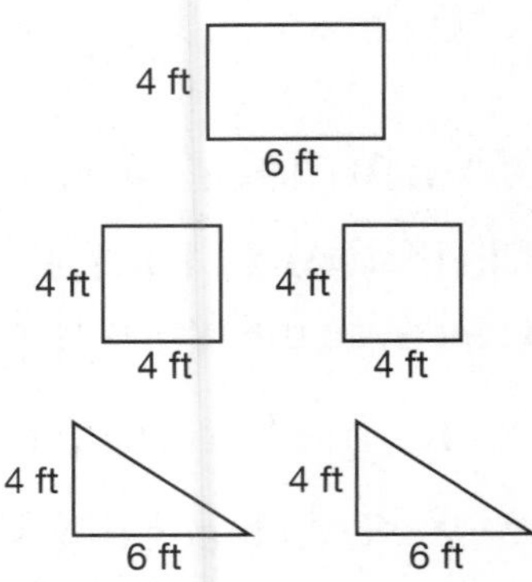

Reread Ask yourself questions as you read.

- What is the problem about?

____

- What will Kim use to make the rug?

____

- What question is the problem asking?

____

Search for Information

Look at the problem and the drawing. Mark the information you will need.

Record What details will help you solve the problem?

Kim has one 4-foot by 6-foot _______________.

She has two 4-foot by 4-foot _______________.

She has two right triangles. Each triangle has one side that measures ________ feet and another side that measures ________ feet.

A strategy that helps you see the pieces of the rug can help you solve this problem.

Decide What to Do

You know the shapes of Kim's rug pieces. You know how large the rug must be.

Ask How can I find out how to make the shapes into a 10-foot by 8-foot rectangle?

- I can use the strategy *Draw a Picture.*
- I can draw an outline of the big rug. Then I can draw in all of the pieces to see how they can fit.

Use Your Ideas

Step 1 On a grid, draw a rectangle that is 10 squares by 8 squares.

Step 2 Draw the largest piece of rug first. That is the 4-foot by 6-foot ________________.

Step 3 Draw one of the triangles.

Step 4 Draw the other triangle.

Step 5 Draw the pieces that are left. Those are the ________________.

The largest piece can be hard to fit after the other pieces are in place.

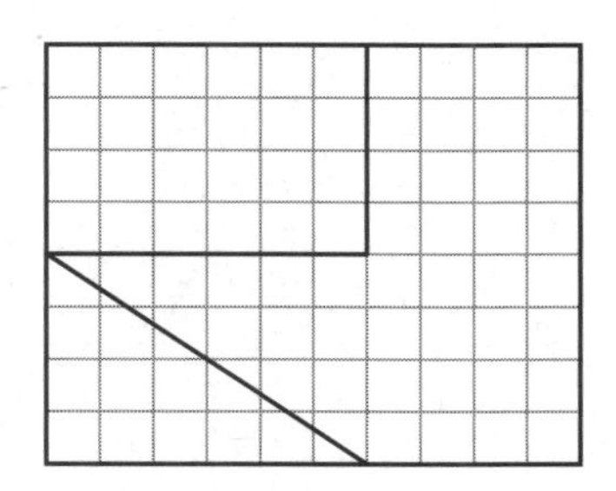

Review Your Work

Check that the shapes you drew fill the outline.

Explain Does the largest piece have to go in a corner? Tell why or why not.

____

____

Try It

Solve the problem.

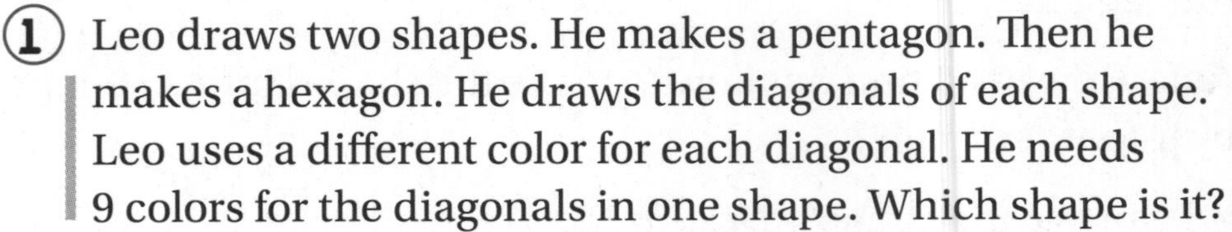

① Leo draws two shapes. He makes a pentagon. Then he makes a hexagon. He draws the diagonals of each shape. Leo uses a different color for each diagonal. He needs 9 colors for the diagonals in one shape. Which shape is it?

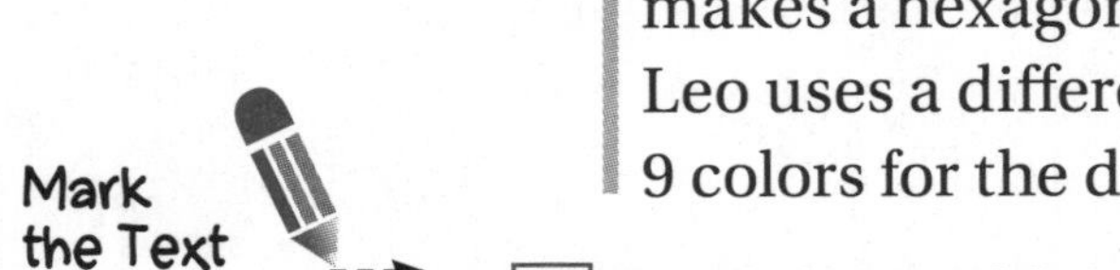

Read the Problem and Search for Information

Reread slowly. Mark any words you do not understand.

Decide What to Do and Use Your Ideas

You know each diagonal is a different color. You can draw each polygon. You can draw the diagonals. Then you can count them.

Step 1 Draw a pentagon. Then draw a hexagon.

Step 2 Draw all the diagonals in each shape.

Step 3 Count the diagonals.

A pentagon has ________ diagonals.

A hexagon has ________ diagonals.

Ask Yourself

Did I draw all the possible diagonals?

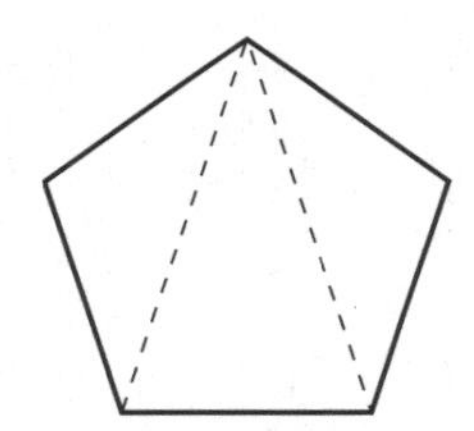

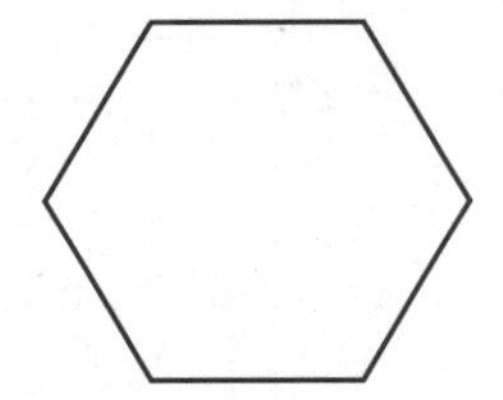

The shape with 9 diagonals is a ________________.

Review Your Work

Be sure you count each diagonal only once.

Conclude How could using a different color for each diagonal help you answer the question?

____

____

Apply Your Skills

Solve the problems.

(2) Dawn starts a design. She draws a hexagon with 3-inch sides. Then she cuts the hexagon into triangles. Each triangle will be the same size. So far, Dawn has cut out 2 triangles. How many triangles will she cut out in all?

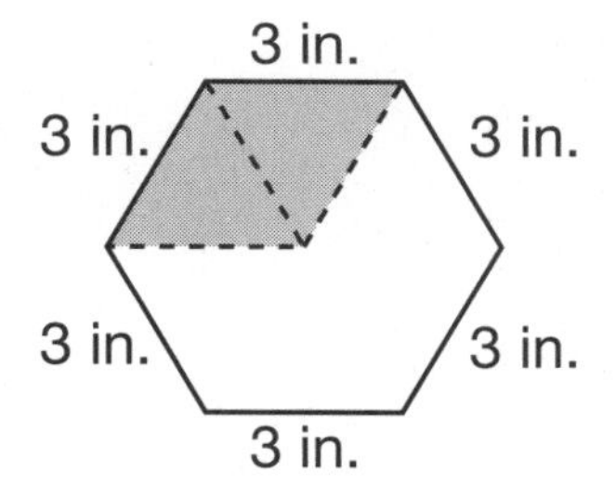

Answer ______________________________

Determine Ming says 4 triangles is the answer. Is that correct? If not, explain how Ming might have gotten that answer.

____

____

Hint You can draw the triangles and count them.

Ask Yourself

Does my drawing match the information in the problem?

(3) The sign for Mr. Kofi's shop window is a 25-inch by 11-inch rectangle. The shop is having a sale. The sale sign is a 25-inch by 8-inch rectangle. Mr. Kofi makes one large rectangular sign by putting the sale sign right above the shop sign. He will put lights around the larger sign. How many inches of lights are needed?

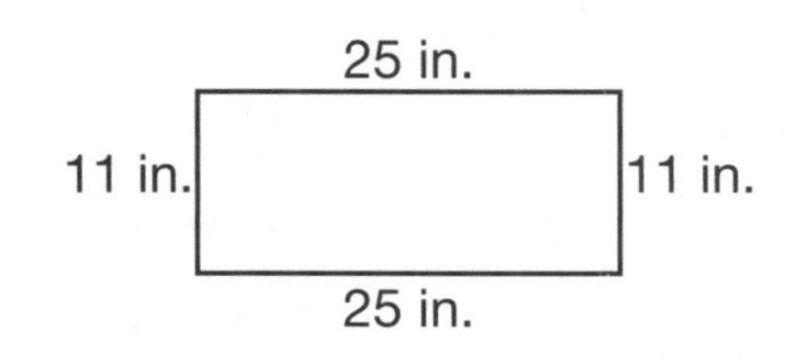

Answer ______________________________

State Which words in the problem tell you how to draw the larger sign?

____

____

Hint Draw the larger sign and label its sides.

Ask Yourself

Which operation will I use to answer the question?

4 Mr. Green is making a sign. The sign will be a rectangle. The rectangle needs to cover 12 square inches. The lengths of the sides are whole numbers. Mr. Green plans to put wood strips around the edges of the sign. He wants to use the least amount of wood strips. What should the lengths of the sides of the sign be?

Hint Draw the rectangles. Label the sides.

Ask Yourself

What is the distance around each rectangle?

The lengths of the sides of the sign could be:

1 inch by ________ inches   2 inches by ________ inches

3 inches by ________ inches

Answer ______________________________

Analyze How does drawing a picture help you?

____

____

5 Larry has a square garden. Each side measures 15 meters. Larry adds space to one side of his garden to make it bigger. That space is shaped like an equilateral triangle. He wants to put a fence that is 1 meter high around the whole garden. How many meters of fence does he need?

15 m

Hint The triangle will share 1 side with the square.

Ask Yourself

How long will the sides of the triangle be?

Answer ______________________________

Identify What information in the problem is *not* needed?

____

____

On Your Own

Solve the problems. Show your work.

6. Sal made a card in the shape of a pentagon. He drew all the diagonals from one corner. Then he cut on the diagonal lines to make 3 triangles. Nora makes a card from a different shape. She draws all the diagonals from one corner. She cuts on the diagonal lines to make 4 triangles. What shape is Nora's card?

Answer ______________________________

Generalize Write a rule to find the shape when you know how many triangles you want to get. Give an example.

____

____

7. Jack is using these 6 tiles to make a tray. The tray will be an 8-inch by 3-inch rectangle. Four of his tiles are right triangles. Find two ways Jack can arrange the tiles to make his tray.

4 in. × 2 in. rectangle; 8 in. × 1 in. rectangle; four right triangles, each 2 in. by 2 in.

Answer ______________________________

Justify Does it make sense to place the 1-inch by 8-inch rectangle tile first? Explain why or why not.

____

____

Create

A sports store puts small squares on baseball caps. Write a problem about making this design. Your problem should be one that can be solved by drawing a picture. Solve your problem.

Strategy Focus

Act It Out

MATH FOCUS: Polygons and Congruence

Learn About It

Bianca uses pattern blocks to make designs. She wants to make hexagons without using the yellow hexagon block. So she uses two red trapezoids to make a hexagon. What are four other ways Bianca can use pattern blocks to make hexagons?

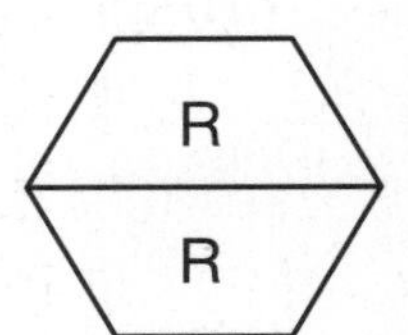

Reread Ask yourself these questions.

- What is this problem about?

____

- What do I know about the designs Bianca wants to make?

____

- What am I asked to find?

____

Search for Information

Read the problem again. Look at Bianca's design.

Record Write what you know about the problem.

You know that Bianca has used two ____________________ to make a hexagon.

You know that you must find four other ways to make ____________________ .

A strategy that lets you build shapes can help you solve the problem.

Decide What to Do

You know you need to make hexagons. You know that you can use different pattern blocks.

Ask How can I find four other ways to make a hexagon with pattern blocks?

- I can use the strategy *Act It Out.* I can use pattern blocks. I can start with a yellow hexagon block. I can try to cover all of it with other pattern blocks.

Use Your Ideas

Step 1 Start with a yellow hexagon pattern block. Bianca already used red trapezoids so choose blue rhombuses. Find out how many will completely cover the hexagon.

Step 2 Write the name and how many pattern blocks you use.
_________ blue rhombuses.

Step 3 Try using only orange squares or only green triangles. Then try using more than one kind of pattern block at the same time.

Not all pattern blocks will make hexagons.

Four other ways Bianca can make hexagons with pattern blocks are to use:

1. ______________________________

2. ______________________________

3. ______________________________

4. ______________________________

Review Your Work

Check that all your ways are different.

Describe Tell how using pattern blocks helps you solve this problem.

____

____

Try It

Solve the problem.

① Ben has a yellow hexagon pattern block. He wants to use it with some other pattern blocks to make a bigger hexagon. What are two ways Ben can make a bigger hexagon using other pattern blocks?

Read the Problem and Search for Information

Reread the problem. Find the details you need.

Decide What to Do and Use Your Ideas

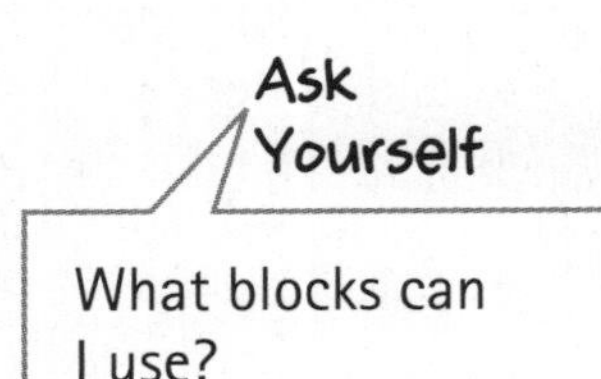

You can act it out. You can put pattern blocks together in different ways.

Step 1 Start with a yellow hexagon. Place a red trapezoid along one side.

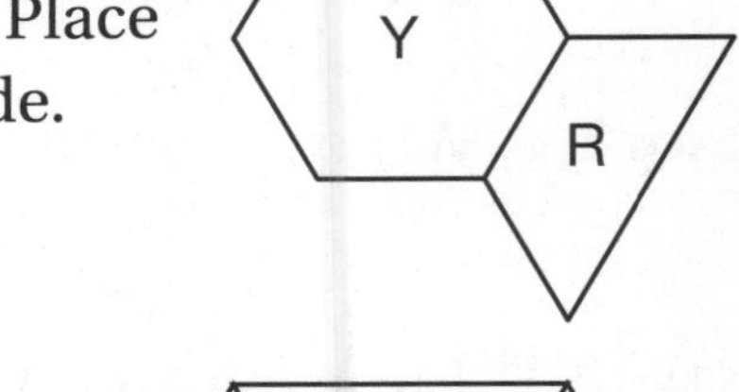

Step 2 Put red trapezoids along the rest of the sides.

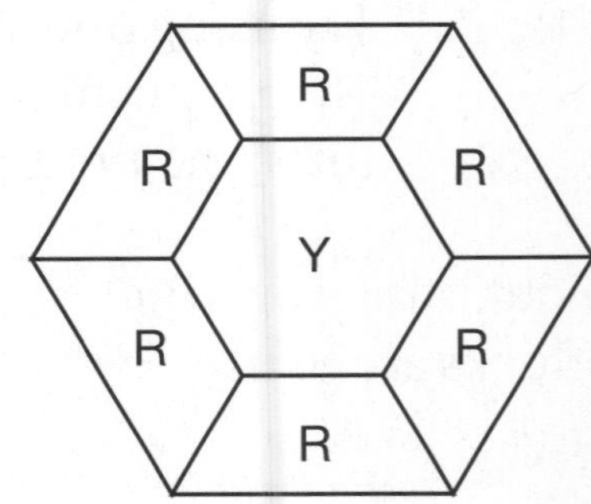

Step 3 Try using more than one kind of block.

Two ways Ben can make bigger hexagons are

1. ____________________

2. ____________________

Review Your Work

Check that each shape you made has 6 sides.

Demonstrate How can you use pattern blocks to make an even bigger hexagon?

____

____

Apply Your Skills

Solve the problems.

② Kwaku plans to use pattern blocks to make a bigger trapezoid. Draw one way that he can do it. Which blocks can Kwaku use?

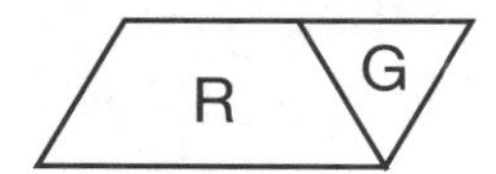

Hint Think about where to put more blocks.

Start with 1 red trapezoid and 1 green triangle.

Ask Yourself

How many of these shapes do I need?

Answer ____________________

Explain Why is the strategy *Act It Out* helpful for solving this problem?

____

____

③ Terry made this design. She wants to make it with other pattern blocks. Terry thinks she can make it in at least four other ways. Is she right? If so, list the blocks Terry can use for the four ways.

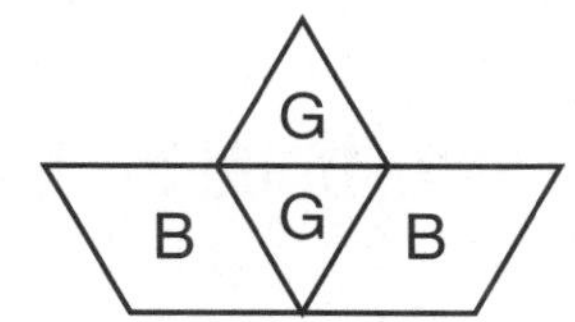

Hint Build Terry's first design using pattern blocks.

Terry uses 2 ____________ and 2 ____________ to make her first design.

Ask Yourself

Are there any blocks that will not work?

Answer ____________________

Summarize What steps did you use to solve this problem?

____

____

Hint If you can move a shape to fit exactly over another shape, then the shapes are congruent.

Ask Yourself

Which shapes look alike?

4. Gil and his friends make these shapes. They each use 5 square pattern blocks. They want all the shapes to be different. Gil thinks some shapes are congruent. Is Gil correct? Explain your answer.

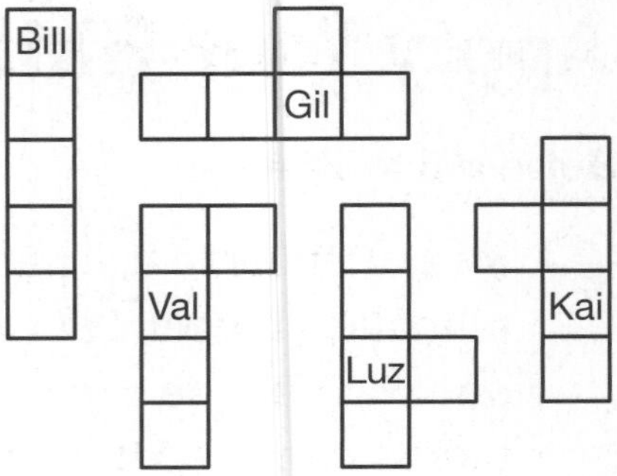

Bill
Gil
Val
Luz
Kai

Use pattern blocks to make each shape if you need to.

Answer ______________________________

Analyze How might tracing the shapes help you answer the question?

____

____

Hint Congruent shapes have exactly the same shape and size.

Ask Yourself

What pattern blocks did Haley use? What other blocks will cover her shape?

5. Haley uses pattern blocks to design this hat. Haley wants to make a congruent shape. She will use a different group of pattern blocks. What blocks can Haley use? Make a drawing to show another way to make the hat.

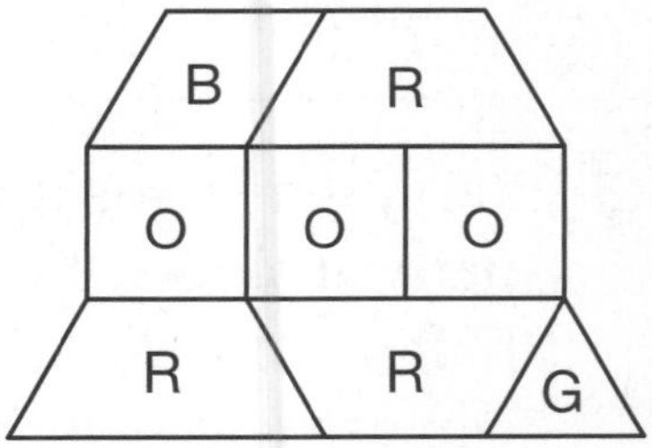

B R
O O O
R R G

You can use pattern blocks to build the hat.

Answer ______________________________

Determine Is there more than one solution to this problem? Explain.

____

____

On Your Own

Solve the problems. Show your work.

6. Abby used pattern blocks to make this shape. She used an orange square, a yellow hexagon, and a red trapezoid. She wants to make a congruent shape with a different group of blocks. What are the most blocks Abby could use? Which blocks would she use?

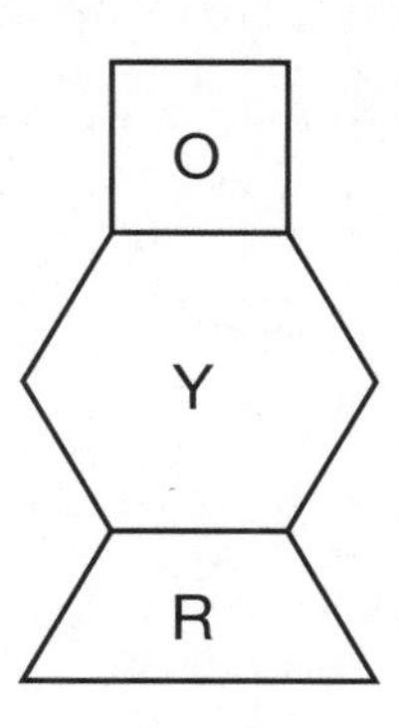

Answer ______________________________

Generalize You want to use the most pattern blocks to make a shape. Which blocks should you use first?

____

____

7. Josh is making posters. He made a design for them. It is made with 10 pattern blocks. How can Josh make a congruent shape with fewer pattern blocks?

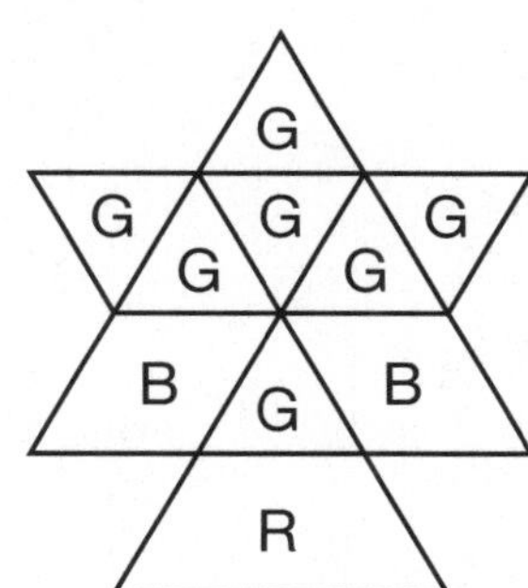

Answer ______________________________

Adapt What other question can you ask about the design?

____

____

Create

Write a problem about making congruent shapes with pattern blocks. Your problem should be one that can be solved using *Act It Out*. Solve it. Draw a picture to show your solution.

Lesson 19

Strategy Focus

Guess and Check

MATH FOCUS: Transformations and Symmetry

Learn About It

Read the Problem

Mr. Cook is putting in a new tile floor. He has one more tile to put in place. He wants to fit the dark gray tile into the open spot with just one move. How should Mr. Cook move the tile?

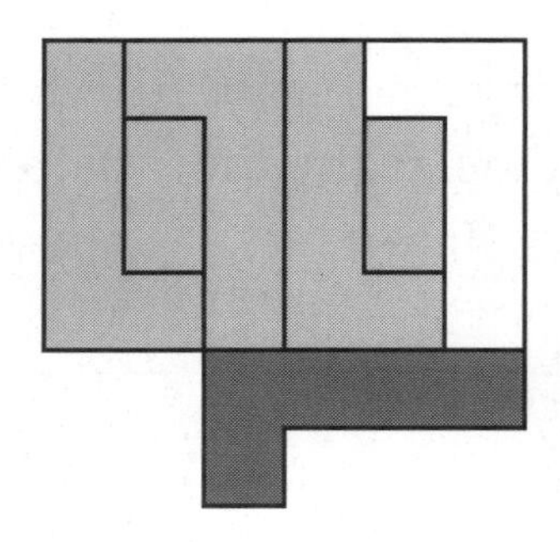

Reread Look at the picture. Then read the problem again. Answer these questions as you read.

- What is Mr. Cook doing?

____

- What question do you need to answer?

____

Search for Information

Mark words or phrases that will help you solve the problem.

Record What details do you know?

Mr. Cook is using ____________________ for the floor.

Mr. Cook needs to fit _________ more tile in the floor.

You can use these details to choose a strategy.

Decide What to Do

You know that Mr. Cook has 1 more tile to put in. You know he wants to use just one move to fit the tile into place.

Ask How can I find how Mr. Cook should move the tile?

- I can use the strategy *Guess and Check*.
- I can move the tile until it fits into the open spot.

Use Your Ideas

Step 1 Try flipping the tile.

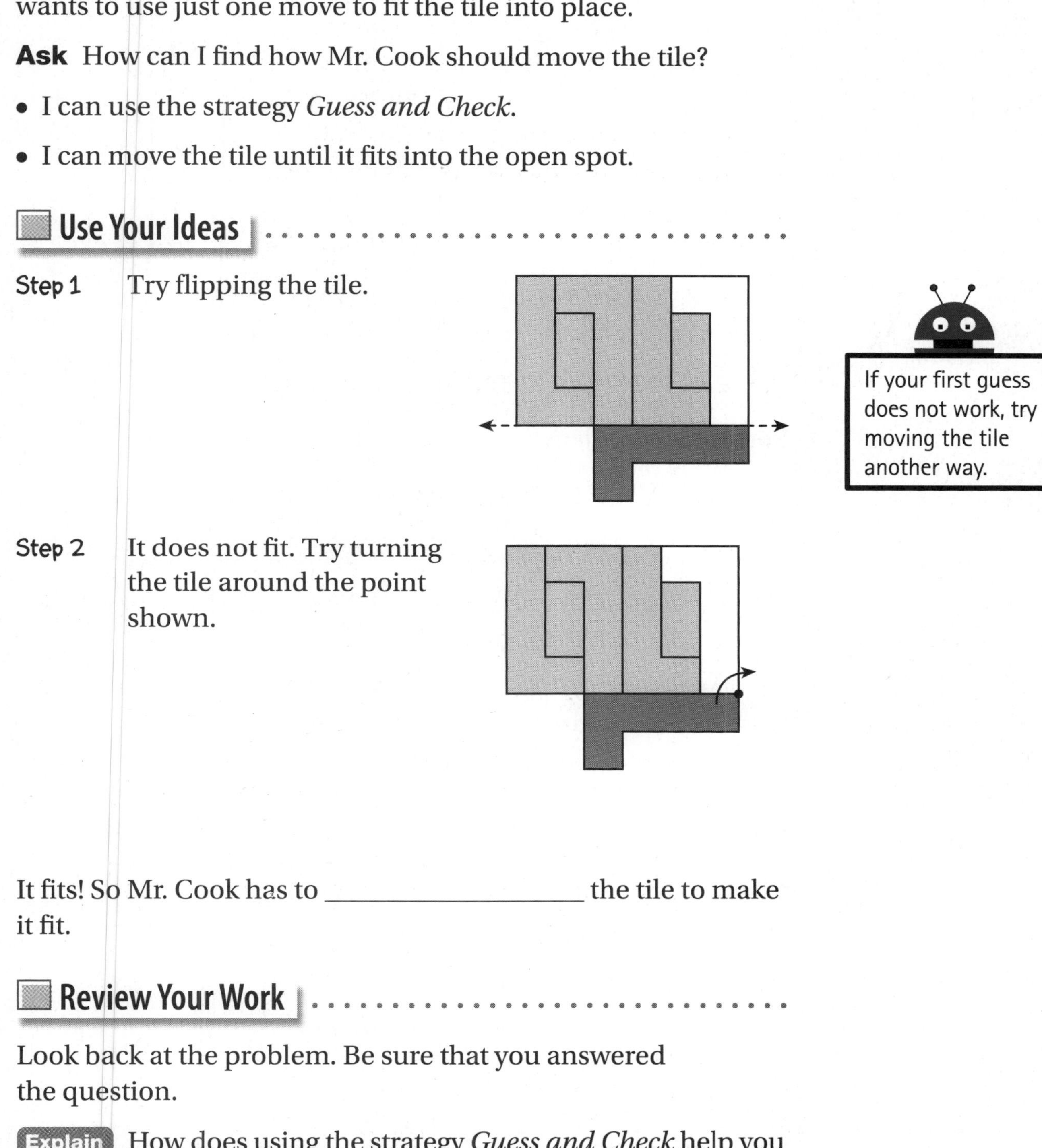

Step 2 It does not fit. Try turning the tile around the point shown.

It fits! So Mr. Cook has to ________________ the tile to make it fit.

Review Your Work

Look back at the problem. Be sure that you answered the question.

Explain How does using the strategy *Guess and Check* help you solve the problem?

____

____

Try It

Solve the problem.

1. Alex is making glass tiles to hang in windows. He will put a different word on each tile. He will use words that he can read from both sides of the window. Alex has already made these two tiles. Which words from the list below can Alex use to make more tiles?

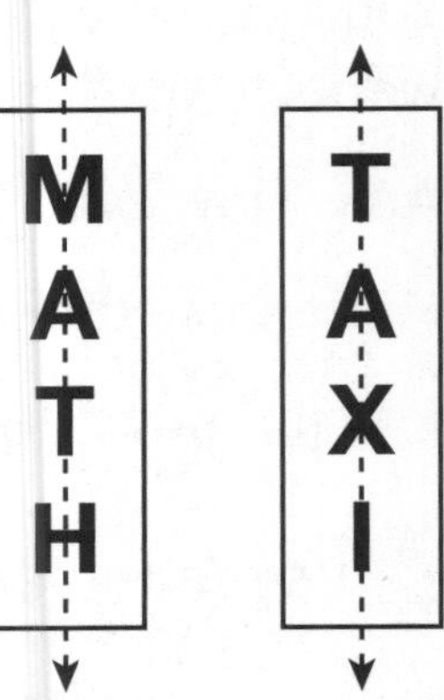

PATH	WHAT	CHAT	WHO
MOP	VOTE	HOW	MAT

Read the Problem and Search for Information

Reread the problem. Decide what the question is.

Decide What to Do and Use Your Ideas

You can use the strategy *Guess and Check.*

Ask Yourself

Do all the letters have a vertical line of symmetry?

Step 1 Start with the word PATH. Stack the letters. Draw a vertical line through the middle.

P A T H

Do the letter parts on both sides match? ______

If not, which letter does not work? __________

Which other word can you cross off the list? Why?

____

Step 2 Try another word. Repeat Step 1.

Step 3 Repeat until you have tried all the words.

Answer Alex can use the words ________________, ________________, ________________, and ________________.

Review Your Work

Look back at your work. Make sure that you tried all the words.

Describe How can the first word you try help you?

____

____

Apply Your Skills

Solve the problems.

② Mari plays a computer game. In this game, puzzle pieces drop down from the top of the screen. Mari gets points when she fills a row. This picture shows a puzzle piece dropping. How should Mari move the piece to fill in the bottom row?

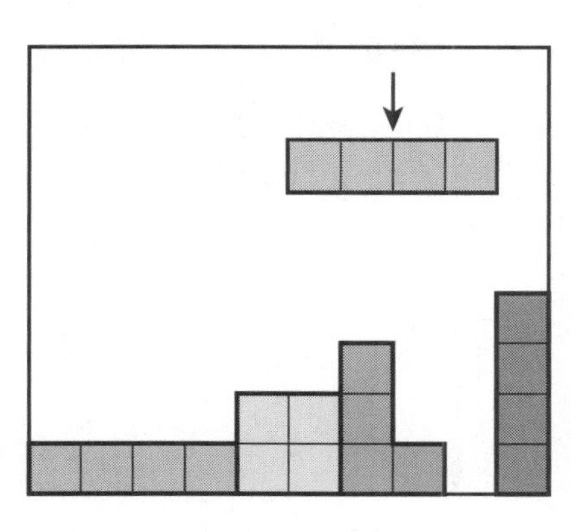

Ask Yourself
How wide is the space where the piece has to fit?

Try sliding the puzzle piece. Try turning the puzzle piece.

Hint Only one move will work.

Answer ______________________________

Predict Suppose the falling piece was a 2 by 2 square. Could you move that piece to fill the bottom row? Explain.

____

____

③ Jean writes in a code. She uses letters with a horizontal line of symmetry. You can see how she writes the word BOX in her code. Which of the words below can she write in code?

Hint A horizontal line goes straight across.

CONE **MEN** **HIDE** **HID**
CANE **TIN** **HEX** **BID**

Draw a horizontal line through the middle of each word.

Ask Yourself
Do both sides match?

Answer ______________________________

Generalize The letter N does *not* have a horizontal line of symmetry. What words in the list can you rule out right away?

____

____

4 Kia is making cards. She uses shapes with a line of symmetry. When she folds them on that line, each half matches exactly. Kia made this heart card and this circle card.

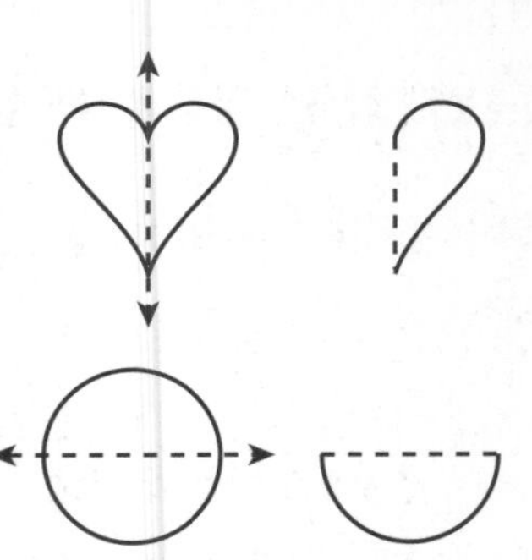

She has the shapes below to make more cards.

Which of these shapes can Kia use for her cards?

Ask Yourself

Will the parts match when I fold the shape along the line?

Try drawing lines of symmetry on the shapes.

Hint Try lines of symmetry that are not horizontal or vertical.

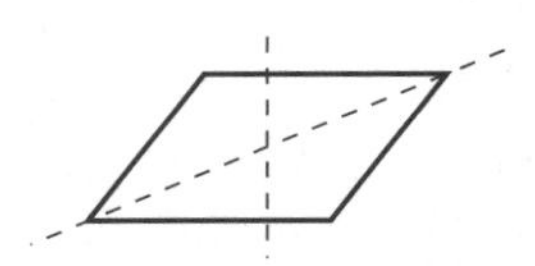

Answer ______________________________

Determine Which shape has more than one line of symmetry? How many does it have?

____

____

5 Mrs. Blair is sewing this quilt. The quilt is a 4-foot by 4-foot square. She has one more piece to put in place. How can Mrs. Blair move the piece to fit it in the quilt?

Hint You can trace the piece.

Ask Yourself

How many ways can Mrs. Blair move the piece?

Try drawing what the piece will look like after a move.

Answer ______________________________

Interpret Does the problem give you any extra information? Explain.

____

____

On Your Own

Solve the problems. Show your work.

(6) Lee is putting together a puzzle. He is looking at puzzle pieces *A*, *B*, and *C*. Which of the three pieces can Lee flip, turn, or slide to fit into the piece shown below? How should he move the piece?

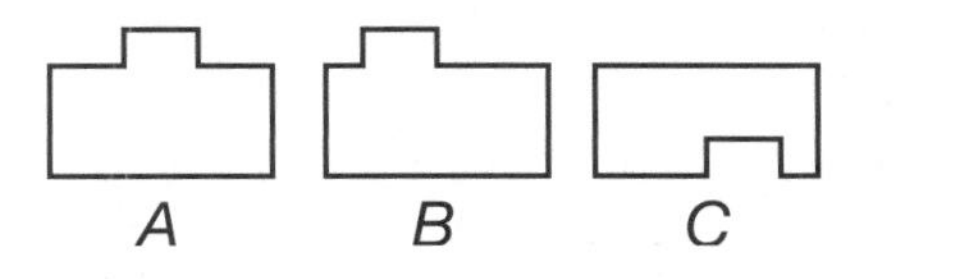

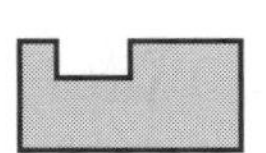

Answer ______________________________

Assess How can you test that your answer is correct?

____

____

(7) Mr. Kang's counter is shaped like a large L. He places one shaded tile in the corner, as shown. He has three more tiles that are the same shape and size. Can he cover his counter exactly with the 4 tiles? Explain.

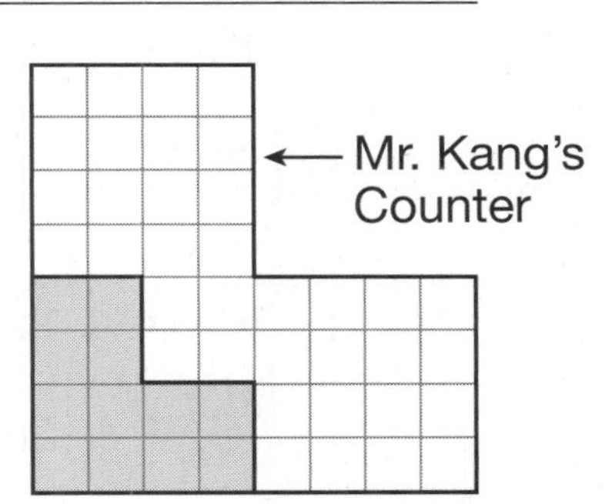

Answer ______________________________

Infer How do you know there *cannot* be any gaps?

____

____

Write a problem about making a design with a shape that you can flip, turn, or slide. Your problem should be one that can be solved by using *Guess and Check*. Show your solution.

Strategy Focus

Make an Organized List

MATH FOCUS: Solid Figures

Learn About It

Read the Problem

Ramon bakes cookies for a bake sale. He wants to make packages for them. The package can be red, blue, or green. Ramon can choose from the shapes at the right. He wants each package to have more than 4 faces. How many different kinds of packages can Ramon make?

Cylinder | Rectangular Prism | Cube

Reread Read the problem again. Ask yourself these questions.

- What does Ramon want to do?

____

- What question do I need to answer?

____

Mark the Text

Search for Information

Reread the problem. Circle words that will help you find the solution.

Record Write what you know about Ramon's choices of colors and shapes.

The colors of the package can be ________________, ________________, ________________.

The shapes Ramon can choose from are ________________, ________________, ________________.

Ramon wants shapes that have more than ________ faces.

You can use these details to choose a problem-solving strategy.

Decide What to Do

You know the colors of the packages. You know the shapes Ramon can choose from. You know that the shape must have more than 4 faces.

Ask How can I find all the packages Ramon can make?

- I can use the strategy *Make an Organized List.*
- I can make a list of the possible colors and shapes.

Use Your Ideas

Step 1 List the possible colors.

red, __________________ , __________________

Step 2 List the shapes that have more than 4 faces.

rectangular prism, __________________

Step 3 Start your list with the first color. Write each color with each of the possible shapes.

Color	Shape
red	rectangular prism
red	cube
blue	rectangular prism
blue	
green	
green	

Making an organized list helps you make sure you do not miss any possibilities.

Ramon can make ________ different packages.

Review Your Work

Check that your answer is what was asked for in the question.

Describe How does making an organized list help you solve the problem?

____

____

Try It

Solve the problem.

1. Ms. Lake makes game pieces out of cylinders, spheres, pyramids, and cubes. Each player needs 2 game pieces. The first piece must be one that can roll. The second piece must be one that cannot roll. How many different pairs of game pieces can she make?

Read the Problem and Search for Information

Reread the problem. Circle words that help you solve it.

Decide What to Do and Use Your Ideas

You can use the strategy *Make an Organized List.*

Step 1 Write the names of the shapes that roll. Write the names of the shapes that cannot roll.

Rolls: cylinder, ________________

Cannot roll: pyramid, ________________

Step 2 List all the possibilities if a cylinder is the first piece. List all the possibilities if a sphere is the first piece.

Ask Yourself

Which shapes can the first piece be? The second piece?

First Piece	Second Piece
cylinder	pyramid
cylinder	cube
sphere	

So Ms. Lake can make ________ pairs of game pieces.

Review Your Work

Check to see that you have not repeated any pairs in your list.

Recognize What words in the problem tell you that the pair *cylinder* and *sphere* should *not* be in your list?

____

____

Apply Your Skills

Solve the problems.

② Ms. Day uses metal, wood, or plastic to make paperweights. Ms. Day will choose from the shapes shown. She wants them to have more than 4 edges. How many different paperweights can Ms. Day make?

Cylinder

Pyramid

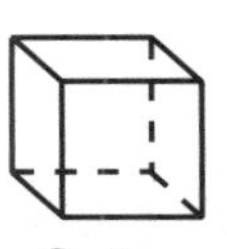

Cube

Shape	Made From
cube	metal
pyramid	metal

Hint Find the shapes that have more than 4 edges.

Ask Yourself

What will the paperweights be made of?

Answer ______________________________

Conclude Why do you find shapes with more than 4 edges first?

____

____

③ Ron has the blocks shown. He uses two blocks to make a clock tower. The block on the bottom cannot roll and does not have a pointed top. The block on top has a pointed top. How many different clock towers can Ron make?

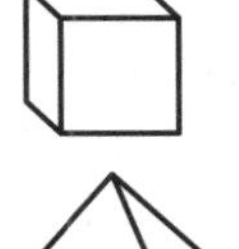

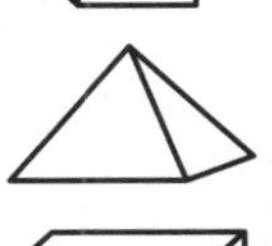

Ask Yourself

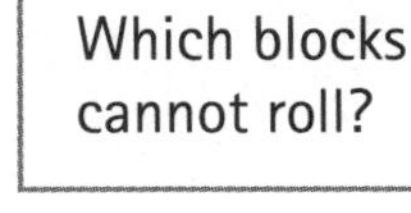

Which blocks cannot roll?

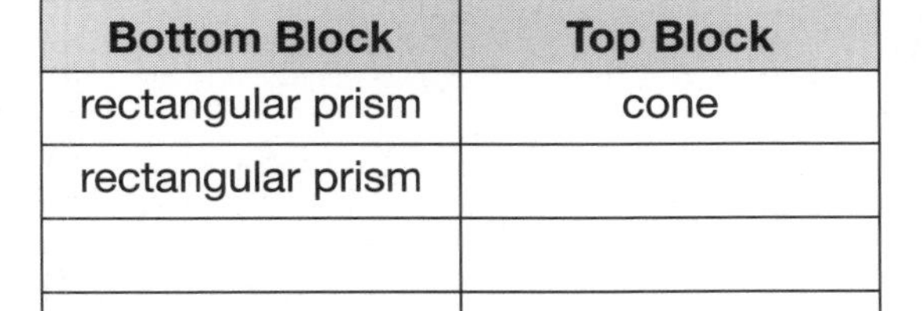

Bottom Block	Top Block
rectangular prism	cone
rectangular prism	

Hint Find the shapes that can go on top.

Answer ______________________________

Explain How did you use the picture to help you solve?

____

____

Ask Yourself

How many different shapes does Sue have?

Hint List each possible picture next to each possible shape.

④ Sue has the shapes below. She will put a picture of a spider, a fly, or an ant on each shape. How many ways can she put the pictures on the shapes?

Shape	Picture
cube	spider

Answer ______________________________

Sequence What steps did you take to solve this problem?

____

____

Ask Yourself

Which shapes have curved surfaces?

Hint Some shapes cannot be paired with some colors.

⑤ Adam's party is in two weeks. He is wrapping favors for his 8 friends. He has boxes shaped like cylinders, cones, and cubes. Boxes with curved surfaces can be blue, silver, or green. The other boxes are always green. Can each friend receive a different box? Explain why or why not.

Make an organized list to show the possible box shapes and colors.

cylinders: blue, __________, __________

________: __________, __________, __________

________: __________

Answer ______________________________

Examine What information is given that you do *not* need?

____

____

On Your Own

Solve the problems. Show your work.

⑥ Venus is making 3-dimensional shapes out of colored paper. She will make cubes and pyramids. Each shape will be red, blue, green, or yellow. How many different shapes can Venus make?

Answer ______________________________

Compare How is this problem like the Learn About It problem?

____

____

⑦ Mr. Ruiz's class is studying 3-dimensional shapes. Each student in his class can choose any one of the shapes shown here. They can write a poem or a song about the shape. How many different choices does each student have?

Answer ______________________________

Modify What if each student could also write a story about the shape? Would your answer change? Explain why or why not.

____

____

Create

Write a problem about making boxes. Make it a problem that can be solved by making an organized list. Use different kinds of shapes and colors. Then solve your problem.

UNIT 5 Review What You Learned

In this unit, you worked with four problem-solving strategies. You can often use more than one strategy to solve a problem. So if a strategy does not seem to be working, try a different one.

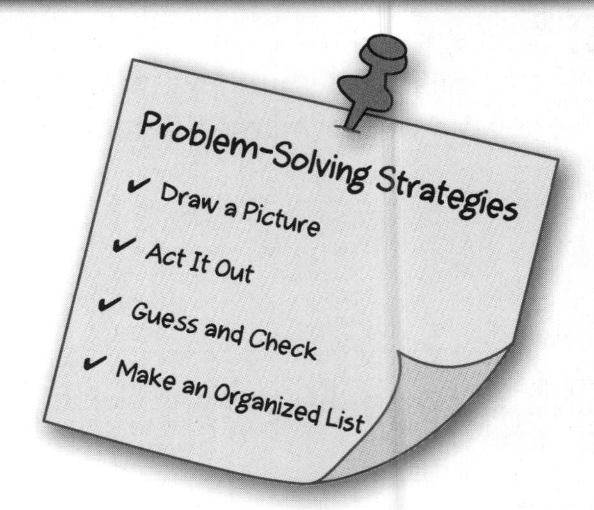

Solve each problem. Show your work. Record the strategy you use.

1. Rita wants to put ribbon around a scarf that she made. The scarf is in the shape of a rectangle. The scarf is 60 inches long and 8 inches wide. How much ribbon will she need?

Answer ______________________

Strategy ______________________

2. Dee is making a design. She begins one design with a hexagon pattern block. She uses other pattern blocks to cover the hexagon. Each of her designs has a line of symmetry. Describe 3 ways she can make her design.

Answer ______________________

Strategy ______________________

3. Look at the picture. How many squares are there? (There are squares of different sizes.)

Answer ______________________________

Strategy ______________________________

4. Sam puts four pattern blocks in a certain order. He puts a triangle block to the left of a hexagon block. He puts a trapezoid block between the triangle and the hexagon blocks. He puts a square block to the left of the triangle block. What is the order of Sam's blocks, from left to right?

Answer ______________________________

Strategy ______________________________

5. Maria puts a fence around her square garden. She puts one post at each corner. Each side has a total of 4 posts. How many posts does she use altogether?

Answer ______________________________

Strategy ______________________________

Explain how drawing a picture helped you solve the problem.

Solve each problem. Show your work. Record the strategy you use.

6. Mrs. Norton has cat treats shaped like the figures shown below. She wants to break them in half along a line of symmetry. Which figures have a line of symmetry?

Answer ______________________________

Strategy ______________________________

7. Look at this puzzle. Which two shapes appear to be congruent?

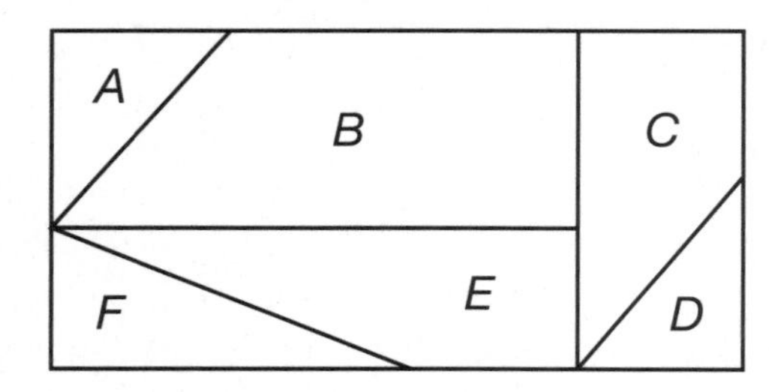

Answer ______________________________

Strategy ______________________________

8. Toshi is choosing a shape for a design. The shape can be a triangle, a rectangle, or a pentagon. The shape can be small, medium, or large. How many possible choices does Toshi have for his shape?

Answer ______________________________

Strategy ______________________________

Explain how you found the different choices. List them.

9. Olivia uses cubes and square pyramids to make a mobile. There are 72 edges in the mobile. How many cubes does she use in the mobile? How many square pyramids does she use?

Answer ______________________

Strategy ______________________

10. A picture has 15 shapes. The shapes are triangles and squares. There are 3 more triangles than squares. How many of each shape are there?

Answer ______________________

Strategy ______________________

Write About It

Think about how you solved Problem 9. What important information is not given in the problem? How did this information help you choose a problem-solving strategy?

____

____

Work Together: Design a Rug

Your group will design a rug for your classroom. Your design must use at least two of the shapes shown. The shapes should fit together exactly with no gaps.

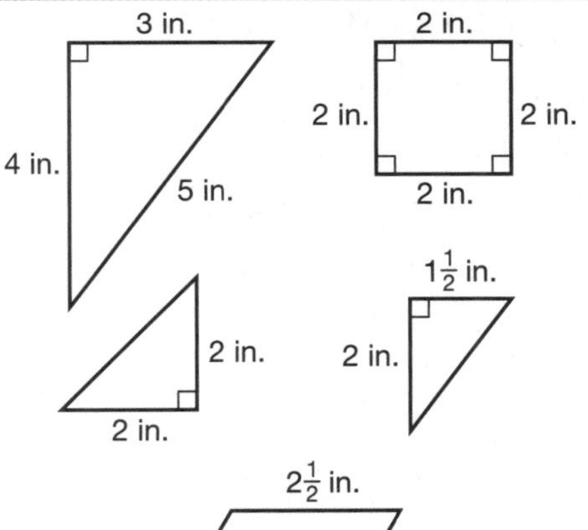

Plan
1. Draw a 6-inch by 8-inch rectangle.
2. Agree on which shapes you will use. Use as many of each shape as you wish.

Arrange The shapes need to completely cover your rectangle. Cut out full size versions of the shapes.

Create Draw and color a final version of your design.

Present As a group, share your model with the class. Explain how you know that the shapes fit together exactly with no gaps and how you made it.

UNIT 6 Problem Solving Using Fractions, Decimals, and Money

Unit Theme:
Daily Life

Math is all around you. It is used in the flower garden at the park. Math is used at the fire station you walk by. It is used in the shops you visit. Math is also used in the bank down the street. In this unit, you will see how you use math in your daily life.

Math to Know

In this unit, you will use these math skills:

- Find parts of a group
- Compare fractions
- Add and subtract decimals and money

Problem-Solving Strategies

- Draw a Picture
- Look for a Pattern
- Make an Organized List
- Work Backward

Link to the Theme

Finish the story. How does Jin tell his dad about the dessert menu? Use some prices in your story.

Jin and his sister are buying dessert for dinner. But they do not know what to buy. So Jin calls home and tells his dad about what the bakery is selling.

____

____

____

____

____

____

____

Use Math Language

Review Vocabulary

Here are some math words for this unit. Knowing what these words mean will help you read the problems.

coin	dime	nickel	penny
denominator	fraction	numerator	quarter

Vocabulary Activity **Math Words**

Some words are found most often in math. Use two words from the list above to complete the sentences.

1. The number below the bar in a fraction is the _________________ .
2. In the fraction $\frac{2}{3}$, the number 3 is the _________________ .
3. The number above the bar in a fraction is the _________________ .
4. In the fraction $\frac{2}{3}$, the number 2 is the _________________ .

Graphic Organizer **Word Web**

Complete the graphic organizer.

- Draw some coins in the top rectangle.
- Use words from the list. Write a word that names a coin in each of the bottom rectangles.

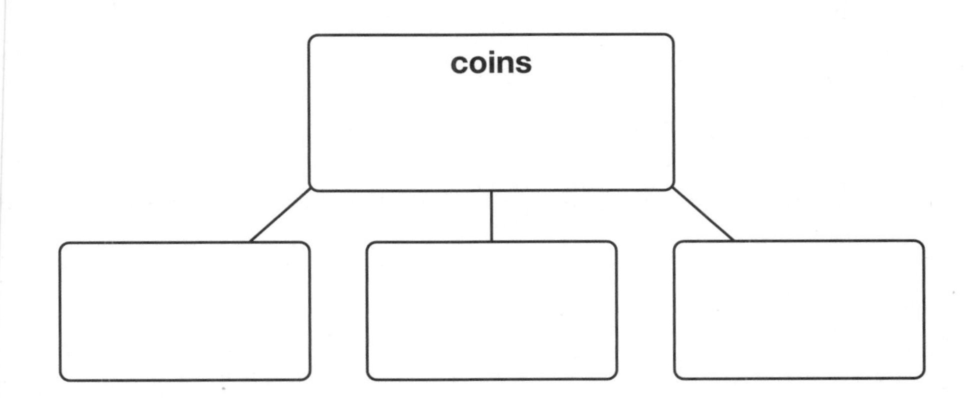

Strategy Focus

Draw a Picture

MATH FOCUS: Fraction Concepts

Learn About It

Read the Problem

The bike rack at the park holds 8 bikes. When Mike gets to the park, the rack is full. He sees that $\frac{1}{4}$ of the bikes are blue. How many bikes are blue?

Reread Think of these questions while you read.

- What is the problem about?

____

- What do you know about the bike rack?

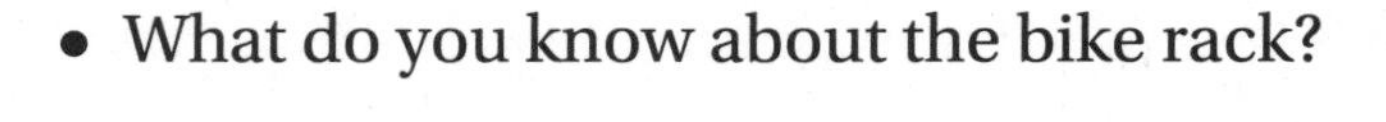

____

- What question do you need to answer?

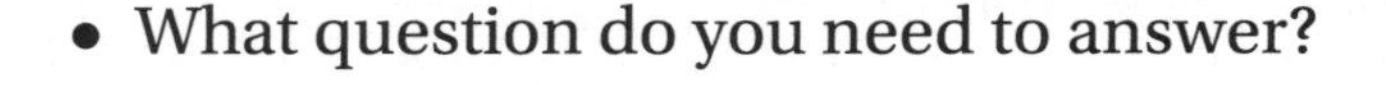

____

Search for Information

Mark words and numbers that will help you.

Record What do you know about the bikes?

There are ________ bikes in the rack in all.

The fraction of the bikes that are blue is ________ .

You can use these facts to choose a problem-solving strategy.

Decide What to Do

You know how many bikes there are in all. You know what part of all the bikes are blue.

Ask How can I find how many bikes are blue?

- I can use the strategy *Draw a Picture* to show the 8 bikes.
- I can put the bikes into equal groups. Then I can find the number of bikes in 1 of the groups.

A drawing can help you picture what is happening in the problem.

Use Your Ideas

Step 1 Draw squares to show 8 bikes.

□ □ □ □ □ □ □ □

Step 2 You want to find $\frac{1}{4}$ of the number of bikes. The denominator of a fraction tells you the number of equal groups. Put the squares in 4 equal groups.

(□ □) (□ □) (□ □) (□ □)

Step 3 The numerator of the fraction tells you how many of the equal groups to consider. Color the squares in 1 of the 4 groups blue.

How many squares did you color blue? ________

So ________ of the bikes are blue.

Review Your Work

Does your picture match the facts in the problem?

Describe How does drawing a picture help you solve the problem?

____

____

Try It

Solve the problem.

1. Three friends will meet at the park. Eva lives $\frac{5}{8}$ mile from the park. Mel lives $\frac{3}{4}$ mile from the park. Tad lives $\frac{1}{2}$ mile from the park. Who lives closest to the park?

Read the Problem and Search for Information

Read the problem again. Circle important words and numbers. Restate the problem in your own words.

Decide What to Do and Use Your Ideas

You can use the strategy *Draw a Picture* to solve the problem. You can use fraction bars to show the fractions.

Step 1 Draw, mark, and color fraction bars to show how far each friend lives from the park.

Ask Yourself

How many parts of each bar should I color?

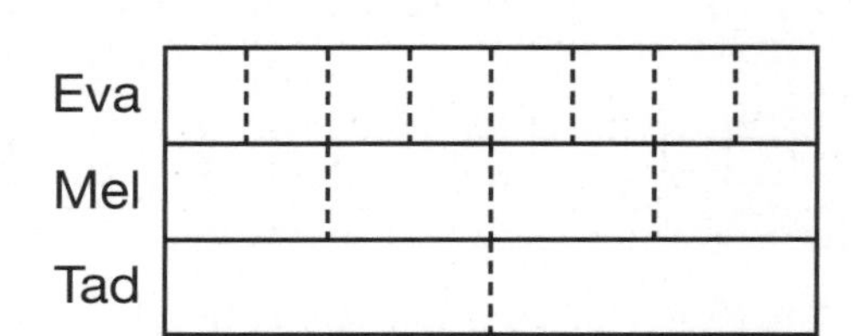

Step 2 Compare the lengths of the bars.

The bar for ________ is the shortest.

So ________ lives closest to the park.

Review Your Work

Did you answer the question the problem asked?

Interpret What information in the problem tells you how to divide each bar?

____

____

Apply Your Skills

Solve the problems.

② An outdoor club is flying kites. There are 12 kites in the sky. $\frac{2}{3}$ of the kites are red. How many kites are red?

Hint Put the kites in 3 equal groups to show thirds.

Ask Yourself

How many of the 3 groups are red?

Answer ______________________________

Conclude Jay said there are 4 red kites. What error might Jay have made?

____

____

③ Seth wants to celebrate July 4th. He plants flowers in his garden. $\frac{1}{4}$ of the flowers are red. $\frac{1}{3}$ of the flowers are blue. Are there more red flowers or more blue flowers?

Red				
Blue				

Hint Use the fraction bars to show $\frac{1}{4}$ and $\frac{1}{3}$.

Ask Yourself

Is the bar for $\frac{1}{4}$ or $\frac{1}{3}$ longer?

Answer ______________________________

Summarize Explain the steps you used to solve this problem.

____

____

4 Bart and James are at the snack bar. They each buy a pizza that costs \$8. The pizzas are the same size. Bart eats $\frac{3}{4}$ of his pizza. James eats $\frac{3}{8}$ of his pizza. Who eats more pizza?

Hint Color each circle to show the fractions.

Bart   James

Ask Yourself

How many parts should I color?

Answer ____________________

Determine Why do you need two drawings?

____

____

5 Mona finds some coins. $\frac{2}{5}$ of the coins are dimes. $\frac{1}{4}$ of the coins are nickels. $\frac{3}{10}$ of the coins are pennies. The rest of the coins are quarters. Does Mona find more dimes or more pennies?

Hint Mark one bar to show fifths. Mark one bar to show tenths. Label the bars.

Dimes

Pennies

Ask Yourself

Which fractions do you need to compare?

Answer ____________________

Select What information is *not* needed to solve this problem?

____

____

On Your Own

Solve the problems. Show your work.

6. Jay's class is going on a picnic. There are 18 picnic tables. The class uses $\frac{5}{6}$ of the tables. How many tables does the class use?

Answer ______________________

Relate How could you make a drawing to solve the problem?

____

____

7. Steve goes to the park at noon on Saturday. He finds these beetles. He wants to put the beetles in order from shortest to longest. What order should Steve put them in?

Beetle	Length (inches)
Firefly	$\frac{5}{8}$
Ladybug	$\frac{1}{4}$
Stink bug	$\frac{3}{8}$

Answer ______________________

Compare How is this problem like Problem 1?

____

____

Create

Look back at the problems in this lesson for ideas. Write and solve your own problem that uses fractions. Your problem should be one where you can draw a picture to solve it.

Strategy Focus

Look for a Pattern

MATH FOCUS: Decimals and Money

Learn About It

Read the Problem

Jan earns money for doing chores. On Day 1, she earned \$1.25. By the end of Day 2, she earned \$2.50. By the end of Day 3, she earned \$3.75. By the end of Day 4, she earned \$5.00. If this pattern continues, how much money will Jan earn by the end of Day 5?

Reread Ask yourself these questions as you read the problem again.

- What is the problem about?

____

- What question must you answer?

____

Mark the Text

Search for Information

Read the problem again. Mark the information you will need.

Record Write the amounts of money Jan earned at the end of each day.

On Day 1, Jan earned ____________________.

By the end of Day 2, Jan earned ____________________.

By the end of Day 3, Jan earned ____________________.

By the end of Day 4, Jan earned ____________________.

You can use this information to help you solve the problem.

Decide What to Do

You know how much money Jan earned so far. You know that the amount she earns follows a pattern.

Ask How can I find how much money Jan will earn by the end of Day 5?

- I can use the strategy *Look for a Pattern* to see how the money Jan earns changes each day.

Use Your Ideas

Day	1	2	3	4	5
Money	$1.25	$2.50	$3.75	$5.00	

_____ _____ _____ _____

The money Jan earned increases each day. Look for an addition pattern.

Step 1 Find the change in the amount from Day 1 to Day 2.

$1.25 + ____________ = $2.50

Write the change under the arrow between Day 1 and Day 2.

Step 2 Find the changes in the amounts from Day 2 to Day 3 and from Day 3 to Day 4. Write a rule.

$2.50 + ____________ = $3.75

$3.75 + ____________ = $5.00

The rule is add ____________.

Step 3 Use the rule to find the amount for Day 5.

$5.00 + ____________ = ____________

So Jan will earn ____________ by the end of Day 5.

Review Your Work

Check that each day follows the rule.

Describe Tell how the strategy *Look for a Pattern* helped you.

____

____

Try It

Solve the problem.

① Fran saved $9.00 to buy treats for her dog. She will use it to buy a bag of treats each day. A bag of treats costs $1.50. How many bags can Fran buy before her money is gone?

Read the Problem and Search for Information

Retell the problem in your own words. Underline the question you need to answer.

Decide What to Do and Use Your Ideas

You can use the strategy *Look for a Pattern* to find how much money Fran has left at the end of each day.

Step 1 Make a table to show the information in the problem. Fran has $9.00 before she buys dog treats.

Step 2 The money Fran has left decreases by the same amount each time.

The rule is subtract ________________.

Ask Yourself

Is this an addition or a subtraction pattern?

Bags	0	1	2				
Money Left	$9.00	$7.50	$6.00				

– $1.50 – $1.50 _____ _____ _____ _____

Step 3 Complete the table. Use the rule. Find how many bags of treats Fran can buy before her money is gone.

So Fran can buy ________ bags of treats.

Review Your Work

Check that you subtracted correctly each time.

Identify How did making a table help you solve the problem?

____

____

Apply Your Skills

Solve the problems.

② Mr. Luz is training for a race. He decides to run farther each week of training. In Week 1, Mr. Luz ran 1 kilometer. In Week 2, he ran 1.5 kilometers. He ran 2 kilometers in Week 3 and 2.5 kilometers in Week 4. If this pattern continues, how far will Mr. Luz run in Week 6?

Ask Yourself

What is the change from Week 1 to Week 2? From Week 2 to Week 3?

Week	1	2	3	4	5	6
Distance (km)	1	1.5	2	2.5		

Hint Check to see if the increase from one week to the next is the same each time.

Answer ______________________________

Summarize What steps did you follow to solve this problem?

____

____

③ The library collects fines for books that are returned late. The chart shows the late fees. Ali's book was 1 week overdue. What will Ali's late fee be?

Days Overdue	Late Fee
1	$0.20
2	$0.40
3	$0.60
4	$0.80

Hint Look for a pattern. See if the late fees change by the same amount.

Day	1	2	3	4	5		
Fee	$0.20	$0.40	$0.60	$0.80			

Ask Yourself

What is the rule?

Answer ______________________________

Explain A book is 2 weeks late. How can you use the rule to find the late fee?

____

____

4 Joe babysits for his younger sister. He gets paid $2.75 for each hour that he babysits. One day, Joe earned $13.75. How many hours did Joe babysit that day?

Hint Joe earns $2.75 more for each hour he babysits.

Hours	1				
Amount	$2.75				

Ask Yourself

Will my answer be in the top row or the bottom row of the table?

Answer ______________________

Determine Which words in the problem tell you whether your answer will be a number or an amount of money?

____

____

Hint Use a place-value table to help you.

5 You and your friend Dana are playing a game called Guess the Pattern. Dana puts out these boxes. Then she adds a box to the right of the one labeled $0.10. What label should be on the new box?

$100	$10	$1	$0.10

Ask Yourself

How are the amounts changing?

The pattern is ______________ by 10.

Answer ______________________

Compare How are the numbers in this pattern like the numbers in a place-value chart?

____

____

On Your Own

Solve the problems. Show your work.

(6) Sue earns money picking up cans. She earns $0.50 on Monday. She earns $1.00 on Tuesday and $1.50 on Wednesday. Sue earns $2.00 on Thursday. How much money will Sue earn on Sunday if this pattern continues?

Answer ______________________________

Consider How could you use the strategy *Look for a Pattern* to solve this problem?

____

____

(7) Chen has $96 to spend. He spends $48 of the money in Week 1. He spends $24 in Week 2. He spends $12 in Week 3. He keeps spending money in the same way. How much money does he have left after 5 weeks?

Answer ______________________________

Discuss Explain how you found a rule for this problem.

____

____

Create

Write a problem of your own that can be solved by looking for a pattern. Solve your problem.

Strategy Focus
Make an Organized List

MATH FOCUS: Money

Learn About It

Read the Problem

Marti has 2 quarters, 5 dimes, 2 nickels, and 5 pennies. She is buying 3 postcards for $1.00. She wants to pay with the greatest number of coins. Which coins will she pay with?

Reread Answer these questions as you read.

- What is the problem about?

____

- What are you trying to find?

____

Mark the Text

Search for Information

Read the problem again. Mark the information you need.

Record Write the coins that Marti has and what she will buy.

She has ________ quarters.

She has ________ dimes.

She has ________ nickels.

She has ________ pennies.

Marti is buying ________ postcards.

She will spend ________ .

You can use these details to choose a problem-solving strategy.

Decide What to Do

You know which coins Marti has. You know you need to find the greatest number of coins that will make $1.00.

Ask How can I find which coins Marti will pay with?

- I can use the strategy *Make an Organized List* to show all the different ways to make $1.00 with the coins that Marti has.
- Then I can see which way uses the greatest number of coins.

Use Your Ideas

Step 1 You can use a table to make an organized list. List all the groups of quarters, dimes, nickels, and pennies that Marti has that will add up to $1.00.

	Quarters	Dimes	Nickels	Pennies	Total
Group 1	2	5	0	0	$1.00
Group 2	2	4	2		$1.00
Group 3	2	4	1		$1.00

Marti has to use the 2 quarters to make the amount she needs.

Step 2 Count the number of coins in each group.

Group 1: 7

Group 2: ________

Group 3: ________

Step 3 Choose the way Marti can pay that uses the greatest number of coins.

Marti will pay with ________ quarters, ________ dimes, ________ nickels, and ________ pennies.

Review Your Work

Check Marti's coins. Did you find all the ways to make $1.00?

Recognize Could Marti use all the coins she has to make $1.00? Why or why not?

____

____

Try It

Solve the problem.

① Jen buys a new hairbrush. She pays with a five-dollar bill. Her change is 13¢. What are all the possible groups of coins that Jen could get back in change?

Read the Problem and Search for Information

Identify what you know and what you need to find.

Decide What to Do and Use Your Ideas

You can use the strategy *Make an Organized List* to find the possible combinations of coins.

Ask Yourself

Will Jen get any quarters back?

Step 1 Use a table to make an organized list. List the possible groups of coins where Jen gets dimes back.

	Dimes	Nickels	Pennies	Total
Group 1	1	0		13¢

Step 2 List the groups where Jen does *not* get any dimes.

Group 2	0	2		13¢
Group 3	0	1		13¢

Step 3 List the groups where Jen does *not* get any dimes or nickels.

Group 4	0	0		13¢

So Jen could get back 1 dime and ________ pennies, 2 nickels and ________ pennies, 1 nickel and ________ pennies, or ________ pennies.

Review Your Work

Check that each group of coins in your list makes 13¢.

Identify What facts are *not* needed to answer the question?

____

____

Apply Your Skills

Solve the problems.

② Mrs. Lapiz pays $1.35 for parking at a meter. She pays with dimes and quarters. How many different ways can she use dimes and quarters to pay for parking?

	Quarters	Dimes	Total
Group 1	5	1	$1.35
Group 2	3		$1.35
Group 3			$1.35

Hint Start your list with 5 quarters. Then find the number of dimes.

Answer ______________________

Examine Why does the list include only odd numbers of quarters?

____

____

Ask Yourself

Can Mrs. Lapiz use only quarters to pay for parking? Can she use only dimes?

③ Berto has $3.50 for lunch. He wants chicken and one other item from this menu. What choices does he have?

Ask Yourself

How do I find how much each lunch costs?

	First Item	Second Item	Total Cost
Choice 1	Chicken	Carrots	$\$2.50 + \$0.75 = \$3.25$
Choice 2	Chicken	Salad	
Choice 3	Chicken		
Choice 4	Chicken		

Hint The total cost is the sum of the cost of both items.

Answer ______________________

Explain How does the table help you make your list?

____

____

④ Greg bought a ball for 88¢. He paid with a one-dollar bill. How many different ways can Greg get his change?

	Dimes	Nickels	Pennies	Total
Group 1	1	0		
Group 2	0			

Ask Yourself

What is the amount of Greg's change?

Hint The problem asks for the number of ways, not which coins Greg will get.

Answer ______________________________

Compare How is this problem like Problem 1? Explain.

____

____

⑤ Dina buys pencils for school. She gets back $0.47 in change. The coins are dimes, nickels, and pennies. She gets the same number of dimes and nickels. She gets fewer pennies than dimes. What coins did Dina get for change?

Hint You know the number of dimes and nickels is the same. Use this information to make an organized list.

	Dimes	Nickels	Pennies	Total
Group 1	1			
Group 2	2			

Ask Yourself

Could Dina get 4 dimes and 4 nickels?

Answer ______________________________

Determine What words in the problem helped you decide which group of coins was the answer to the problem? Explain.

____

____

On Your Own

Solve the problems. Show your work.

6. Hanna has two $1 bills, 6 quarters, 2 dimes, 8 nickels, and 6 pennies. She buys a birthday card for $2.91. She wants to pay the $0.91 with the greatest number of coins possible. Which coins should she pay with?

Answer ______________________________

Decide Did you start your list by using the greatest number of quarters, dimes, nickels, or pennies? Explain.

____

____

7. Bart has $4.50. He wants to buy 1 flower for his sister and 1 flower for his mother. He can buy the same kind or different kinds of flowers. What kinds of flowers can Bart buy?

Answer ______________________________

Conclude How could you use an organized list to help you solve this problem?

____

____

Look back at the problems in the lesson. Write your own problem about money that can be solved by making an organized list.

Strategy Focus

Work Backward

MATH FOCUS: Add and Subtract Decimals and Money

Learn About It

Read the Problem

Coach Brown buys 10 T-shirts for his team. Each T-shirt costs $6. He also buys a soccer ball for $22. Coach Brown has $3 left. How much money did he have at the start?

Reread Read the problem again. Ask yourself these questions.

- What is the problem about?

____

- What kind of information is given?

____

- What do I need to find out?

____

Mark the Text

Search for Information

Read the problem again. Circle the numbers you need.

Record Write what you know.

Coach Brown buys ________ T-shirts.

Each T-shirt costs ________ .

Coach Brown buys ________ soccer ball.

The soccer ball costs ________ .

He has ________ left.

You can use what you know to solve the problem.

Decide What to Do

You know how much money Coach Brown has left. You also know how much he spent.

Ask How can I find how much money Coach Brown had at the start?

- I can use the strategy *Work Backward.*
- I can start with how much Coach Brown has left and work back to the start.

Use Your Ideas

Step 1 Start with the amount Coach Brown has left.

Coach Brown has $3 left.

Step 2 Work backward. Add what he spent on the soccer ball to the amount he has left.

$3 + ________ = $25

10 T-shirts cost $6 each.
10 × $6 = $60

Step 3 Add what he spent on T-shirts to the sum from Step 2. The result will be the amount at the start.

$25 + $60 = ________

So Coach Brown had ________ at the start.

Review Your Work

Check that your answer is correct by working forward.

Summarize When is the strategy *Work Backward* useful for solving a problem?

____

____

Try It

Solve the problem.

① Ed, Tim, and Amy raised money for the park. Ed raised $14.50. Ed raised $5.00 more than Tim. Tim raised $8.00 less than Amy. How much did Amy raise?

Read the Problem and Search for Information

Reread the problem to be sure you understand it. Circle the words that will help you solve the problem.

Decide What to Do and Use Your Ideas

You know how much money Ed raised. You can use the strategy *Work Backward* to find how much money Amy raised.

Step 1 Start with what you know.

Ed raised $14.50.

What operation undoes addition? What operation undoes subtraction?

Step 2 Work backward.

Ed raised $5.00 more than Tim. Subtract $5.00 to find how much Tim raised.

$14.50 − $5.00 = ________

Step 3 Keep working backward.

Tim raised $8.00 less than Amy. Add $8.00 to the amount from Step 2 to find what Amy raised.

$9.50 + ________ = ________

So Amy raised________ .

Review Your Work

Check that you answered the question that was asked.

Define What words in the problem tell you whether to add or subtract to work backward?

____

____

Apply Your Skills

Solve the problems.

② Ross is saving money for the school fair. He starts with dimes in a jar. Then he puts 3 quarters in the jar. Now he has 95¢. How many dimes did Ross have in the jar at the start?

Now Ross has ________ cents in the jar.

He has ________ quarters and some dimes in the jar.

Ask Yourself

What is the value of 3 quarters?

Answer ______________________________

Hint The problem asks for the number of dimes, not the value of the dimes.

Show What steps did you take to solve the problem?

____

____

③ Ester goes to the movies with her family. There are 2 adults and 3 children. After they buy tickets, they have $9 left. How much did Ester's family have at the start?

Movie Tickets	
Adult	$10
Child	$7

Hint Use the table to find the numbers you need.

Ester's family spent ________ on adult tickets.

Ester's family spent ________ on child tickets.

Ask Yourself

What operation can I use to find how much the family spent in all for each kind of ticket?

Answer ______________________________

Identify How did you decide what to use as your start?

____

____

4 Kyle went to a basketball game. He paid $4.00 for a ticket. During the game he bought a drink for $1.25. He also bought popcorn for $3.00. At the end of the game, Kyle had $0.75 left. How much money did Kyle bring to the game?

Ask Yourself

What operations will I use to work backward?

At the end of the game, Kyle had ________ left.

He paid ________ for popcorn.

He paid ________ for a drink.

He paid ________ for a ticket.

Hint Put your answer into the problem. Then work forward to check that it is correct.

Answer ____

Explain When you work backward, what must you do with each action in the problem?

____

____

5 Dena walks every day. She walked the same distance on Day 1 and Day 2. On Day 3, she walked 1 kilometer. On Day 4, she walked 2.25 kilometers. In all, she walked 7.25 kilometers. How far did Dena walk on Day 1?

Hint Start with the distance Dena walked in all.

In all, Dena walked ________ kilometers.

Ask Yourself

I know that Dena walked the same distance on Day 1 and Day 2. How does that help me solve the problem?

Answer ____

Demonstrate Show how you know that your answer is correct.

____

____

On Your Own

Solve the problems. Show your work.

(6) A school buys 5 red balls, 3 white balls, and 2 green balls for field day. The red balls cost 20¢ each. Each red ball costs 5¢ more than a white ball. Each white ball costs 10¢ less than a green ball. What does each green ball cost?

Answer ______________________________

Judge Are there numbers in the problem that you do *not* need? Explain.

____

____

(7) Kiri buys items for a craft show. He buys 3 paintbrushes. The paintbrushes each cost the same amount. Then he buys a set of paints for $8.50. He spends $14.50 altogether. How much does 1 paintbrush cost?

Answer ______________________________

Discuss What strategy besides *Work Backward* could you use to solve this problem?

____

____

Create

Look back at the problems in this lesson. Write and solve your own *Work Backward* problem.

UNIT 6 Review What You Learned

In this unit, you worked with four problem-solving strategies. You can often use more than one strategy to solve a problem. So if a strategy does not seem to be working, try a different one.

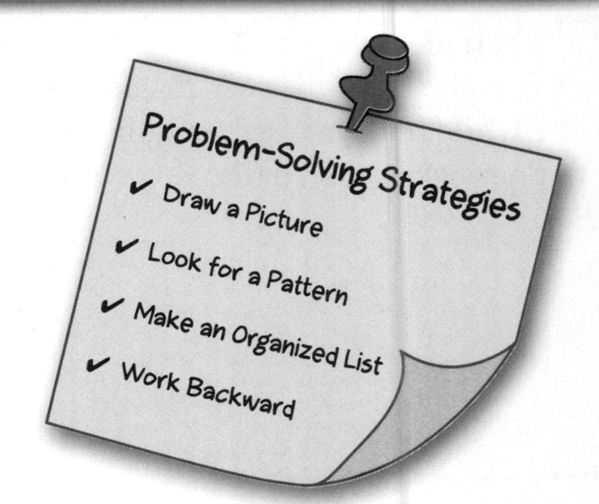

Solve each problem. Show your work. Record the strategy you use.

1. An airplane has 2 rows of seats. Both rows have the same number of seats. People sit in $\frac{2}{6}$ of the seats in Row A. People sit in $\frac{1}{2}$ of the seats in Row B. Which row has more empty seats?

Answer ______________________________

Strategy ______________________________

2. Mrs. Ito buys 3 cookies at the school fair. Each cookie is the same price. She also buys a drink for $1.25. Altogether, Mrs. Ito spends $4.25. How much does each cookie cost?

Answer ______________________________

Strategy ______________________________

3. Lee and Ron are on a sidewalk. The sidewalk is 6.5 meters long. Lee and Ron walk toward each other. Lee walks 1.5 meters and stops. Ron walks 2 meters and stops. How far apart are Lee and Ron when they stop?

Answer ______________________________

Strategy ______________________________

4. Ken is saving money. He has $4.25 in May. He has $6.00 in June. He has $7.75 in July and $9.50 in August. If this pattern continues, how much money will Ken have in September?

Answer ______________________________

Strategy ______________________________

5. Mr. Scott works at the bagel shop. One bagel costs $0.65. Kim gives Mr. Scott 1 dollar to pay for 1 bagel. Mr. Scott gives Kim dimes and nickels for change. How many different ways could Mr. Scott give the change?

Answer ______________________________

Strategy ______________________________

Explain the steps you used to solve this problem.

Solve each problem. Show your work. Record the strategy you use.

6. Mrs. Webb knits scarves. She uses 3.5 spools of thread to make 1 scarf. She uses 7 spools of thread to make 2 scarves. She uses 10.5 spools of thread to make 3 scarves. How many scarves can Mrs. Webb make with 21 spools of thread?

Answer ______________________________

Strategy ______________________________

7. Sofia has 85 cents in her pocket. She spends 69 cents at the store. What are all possible combinations of coins she can have in her pocket now?

Answer ______________________________

Strategy ______________________________

8. Diego's baseball team has $60.00. His team uses the money to buy dinner after each game. The dinner costs the same amount each time. The team has $52.50 left after 1 game. It has $45.00 left after 2 games. How many dinners can the team buy before there is no money left?

Answer ______________________________

Strategy ______________________________

Explain how you found the rule for the pattern in the problem. Write the rule.

9. Nell wants to make a design with strips of colored paper. $\frac{3}{8}$ of the design is blue. $\frac{1}{4}$ of it is green. The rest of the design is yellow. Does Nell need more blue paper or more yellow paper?

Answer ______________________

Strategy ______________________

10. Joe, Hal, and Rod run a race. Each person runs one part of the race. The race is 4 miles long. Joe runs some miles. Hal runs 1.2 miles. Rod runs 1.3 miles. How many miles does Joe run?

Answer ______________________

Strategy ______________________

Write About It

Look back at Problem 2. Explain how you can check your answer.

____

____

Work Together: Design a Banner

Your team is creating a design for a school banner.

Plan Discuss ideas for the design. Talk about how you will use different colors.

Use fractions or decimals to describe each design.

Decide Choose the design your team likes best. Draw your design on a square hundred grid.

Present As a group, show your design to the class. Use fractions and decimals to tell what part of the design each color is.

Math Vocabulary Activities

On the next six pages are some of the math terms you have worked with in each unit.

You can cut these pages to make vocabulary cards. The games and activities below can help you learn and remember the meaning of these important terms.

Try This!

- Complete the activity on the back of each vocabulary card. Use a separate sheet of paper. Discuss your work with a partner or in a small group.
- Work with a partner. Take turns. One person chooses a vocabulary card and shows the front of the card. The other person gives the definition. Check to see if your partner was correct by looking at the back of the card.
- Draw a picture or write an example of each term on a separate card. Then have a partner match your example or picture to a vocabulary card.
- Play a matching game with a partner or a small group.

  Make your own sets of cards. Write a term on one side and leave the other side blank. Lay out the cards in this set in rows, facedown. Make another set for each definition. Write a definition on one side and leave the other side blank. Lay out the cards in this second set in rows, facedown and separate from the first set.

  Take turns. The first player turns over two cards, one from each set. If the cards show a word and its matching definition, the player keeps them and takes another turn. If the word and definition do not match, place the cards facedown where they were, and it is the next player's turn. The player with the most matched pairs wins the game.

Math Vocabulary

Unit 1

addend

$5 + 3 = 8$

addend addend

compare

$2 < 3$

$3 > 2$

difference

$10 - 7 = 3$

difference

estimate

99 + 103 is *about* 200.

number sentence

$3 + 7 = 10$

place value

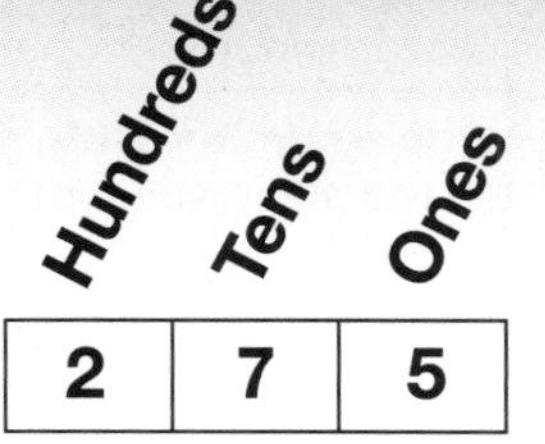

round

17 rounds to 20

sum

$5 + 3 = 8$

sum

Unit 2

array

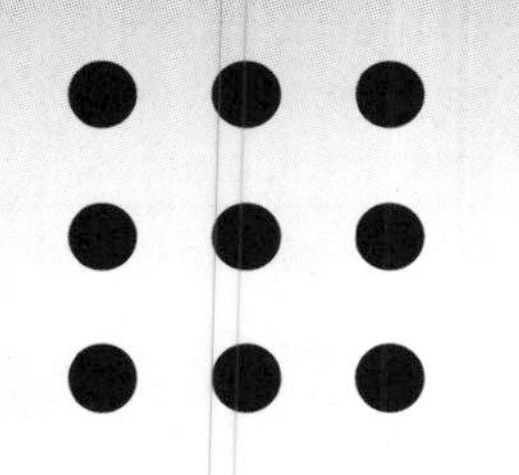

divide

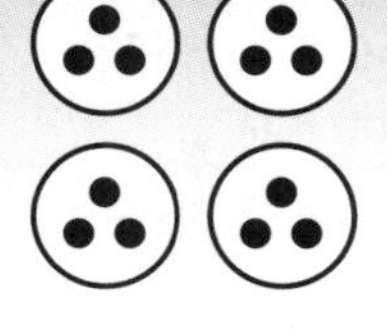

$12 \div 3 = 4$

equal groups

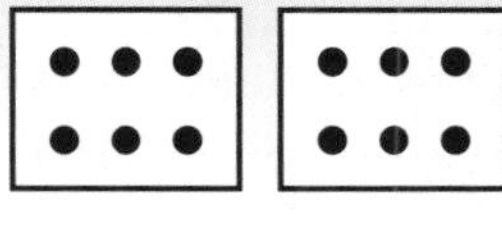

2 groups of 6

factor

$2 \times 3 = 6$

factor factor

multiply

4

2

$2 \times 4 = 8$

pattern

1, 3, 5, 7, . . .

product

$2 \times 3 = 6$

product

rule

2, 4, 6, 8

rule: *add 2*

Math Vocabulary

Unit 1

estimate

a number close to an exact amount

Use *estimate* in a sentence.

difference

the answer in subtraction

Use numbers to show an example of a difference.

compare

to determine which of two numbers is less and which is greater

Write two numbers and compare them.

addend

a number that is added to another number to give a sum

Use numbers to show an addend.

sum

the answer in addition

Use numbers to show a sum.

round

to increase or decrease a number to the nearest given place value

Use *round* in a sentence.

place value

the value of a digit's place in a number

Write a number and tell the place value of each digit.

number sentence

a statement with numbers, at least one operation symbol, and $=$, $<$, or $>$

Give an example of a number sentence.

Unit 2

factor

a number that is multiplied by another number to give a product

Use numbers to show a factor.

equal groups

sets with the same number of objects

Draw a picture that shows equal groups.

divide

to separate into equal groups; to find a quotient

Use *divide* in a sentence.

array

a set of numbers or objects arranged in rows and columns

Draw a picture of an array.

rule

a way to describe a pattern

Write a pattern and a rule for that pattern.

product

the answer in multiplication

Write an example of a product.

pattern

objects or numbers that follow a rule

Draw a picture that shows a pattern.

multiply

to combine equal groups; to find a product

Write two numbers and show how to multiply them.

Math Vocabulary

bar graph

column

Pet Store Sales

Rabbits	12	$20
Frogs	4	$5
Fish	7	$2
Birds	8	$10

line plot

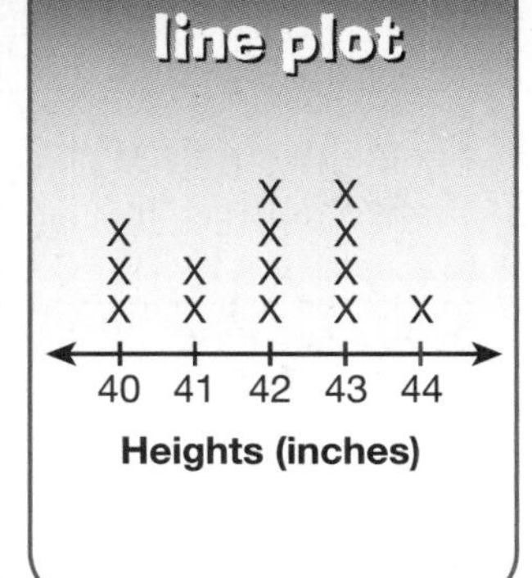

outcome

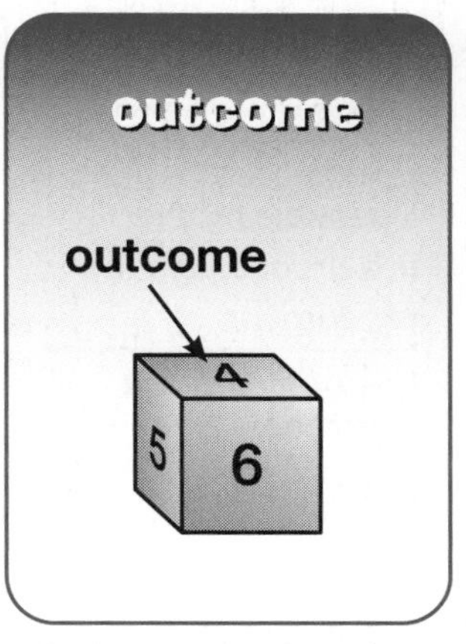

pictograph

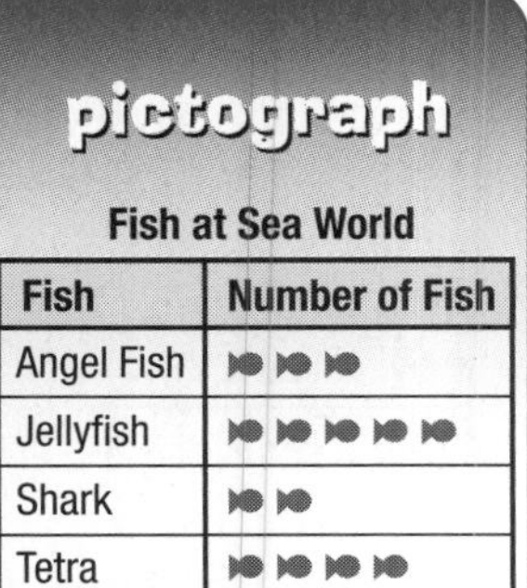

Fish at Sea World

Fish	Number of Fish
Angel Fish	
Jellyfish	
Shark	
Tetra	

Key: Each stands for 5 fish.

probability

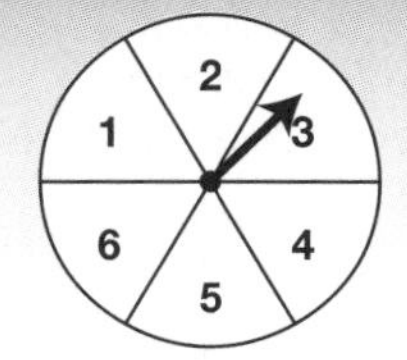

The probability of spinning a 3 is 1 out of 6.

row

Pet Store Sales

Rabbits	12	$20
Frogs	4	$5
Fish	7	$2
Birds	8	$10

row

tally chart

Favorite Pets

Pet	Number
Cat	||||
Dog	~~||||~~ |||
Rabbit	|||

centimeter

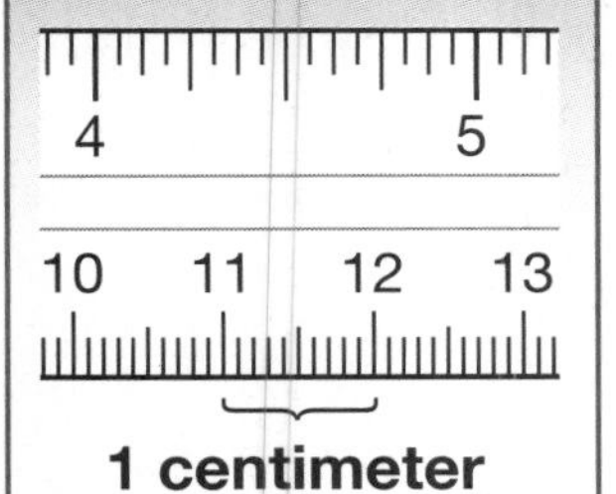

foot

A man's shoe is about 1 foot long.

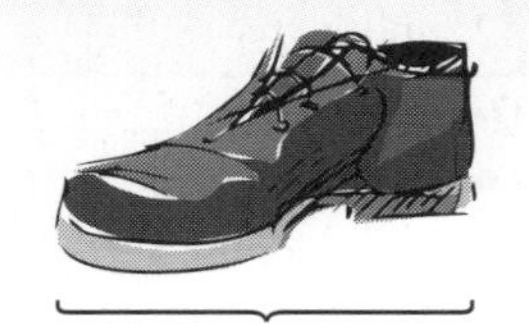

1 foot

inch

length

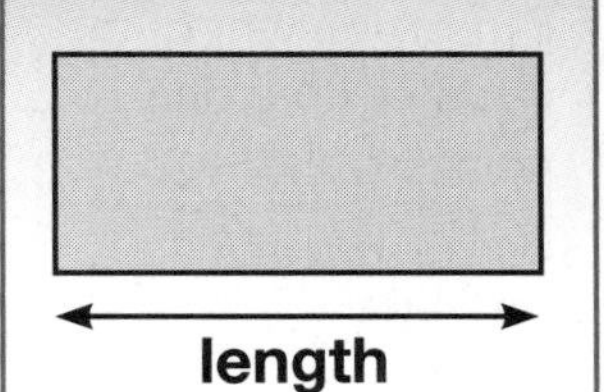

ounce

A slice of bread weighs about 1 ounce.

perimeter

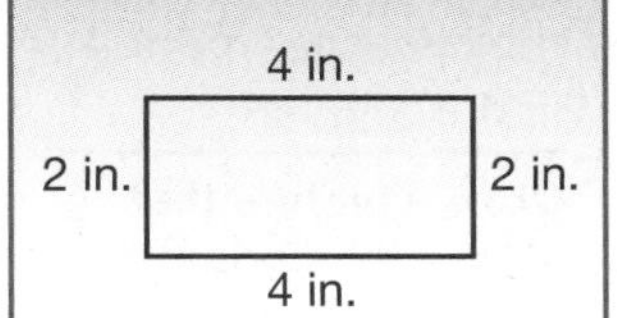

Perimeter = 12 inches

pound

A soccer ball weighs about 1 pound.

volume

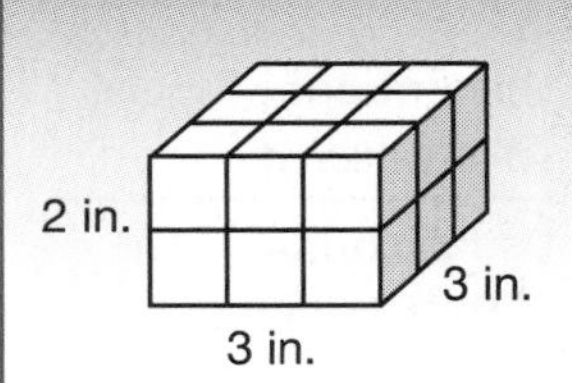

Volume = 18 cubic inches

Math Vocabulary

Unit 3

outcome

a possible result in a probability experiment

Use *outcome* in a sentence.

line plot

a graph that uses Xs above a number line to display data

Draw a line plot.

column

an arrangement of things one on top of another

Draw an example of a column.

bar graph

a graph that uses bars to show data

Draw a bar graph.

tally chart

a chart that uses tally marks to keep track of things being counted

Draw a tally chart.

row

an arrangement of things side by side

Draw an example of a row.

probability

the chance of an event happening

Use *probability* in a sentence.

pictograph

a graph that uses pictures or symbols to show data

Draw a pictograph.

Unit 4

length

a measurement of the distance from one point to another

Use *length* in a sentence.

inch

a customary unit of length

Use *inch* in a sentence.

foot

a customary unit of length equal to 12 inches

Use *foot* in a sentence.

centimeter

a metric unit of length equal to 0.01 meter

Use *centimeter* in a sentence.

volume

the amount of space a solid takes up

Draw a picture that shows volume.

pound

a customary unit of weight equal to 16 ounces

Use *pound* in a sentence.

perimeter

the distance around a figure

Draw a picture that shows the perimeter of a shape.

ounce

a customary unit of weight

Use *ounce* in a sentence.

Math Vocabulary

Unit 5

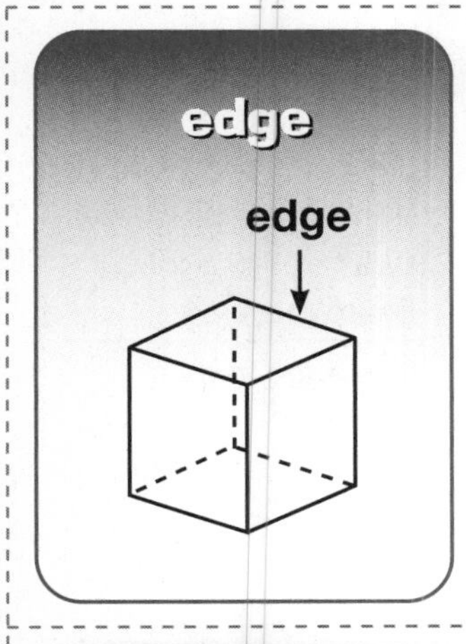

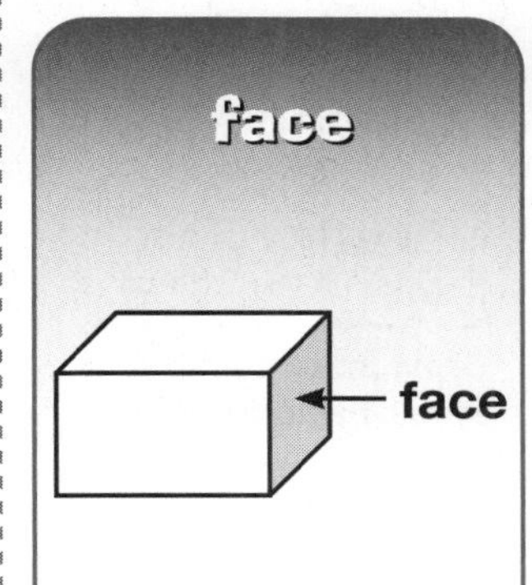

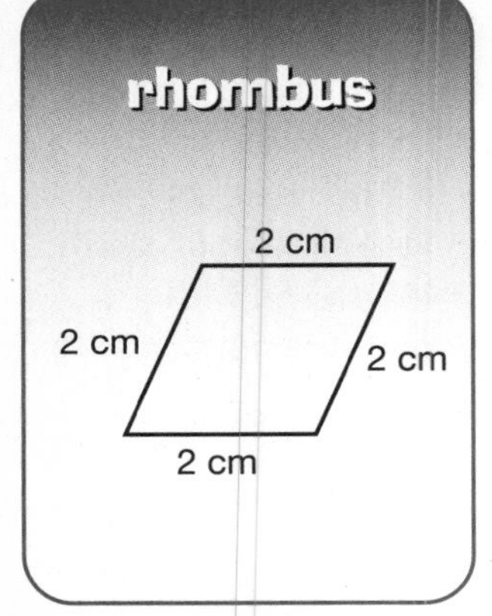

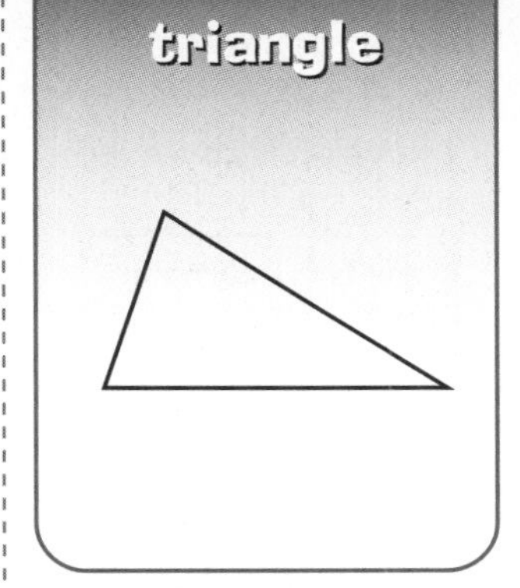

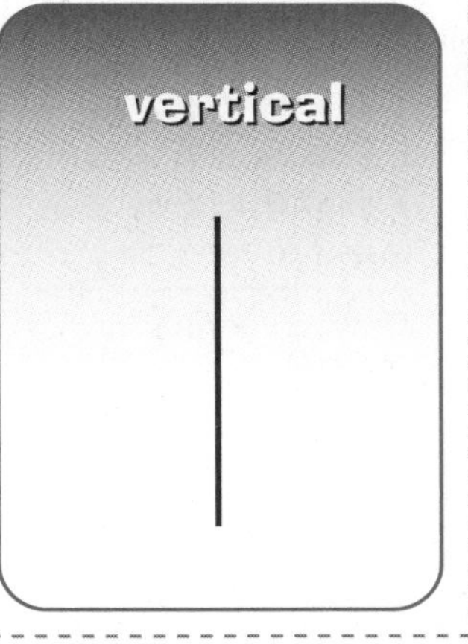

Unit 6

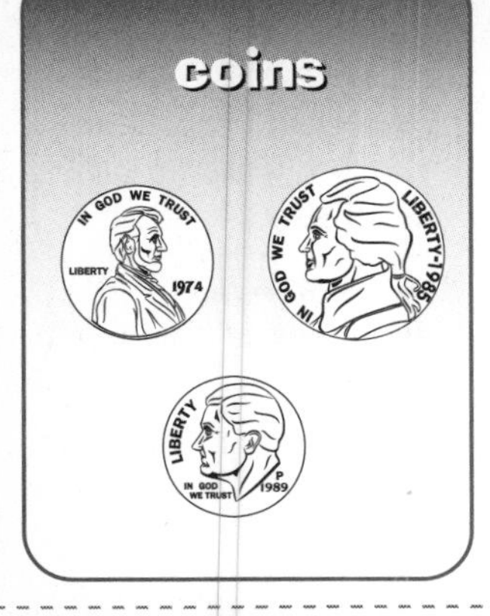

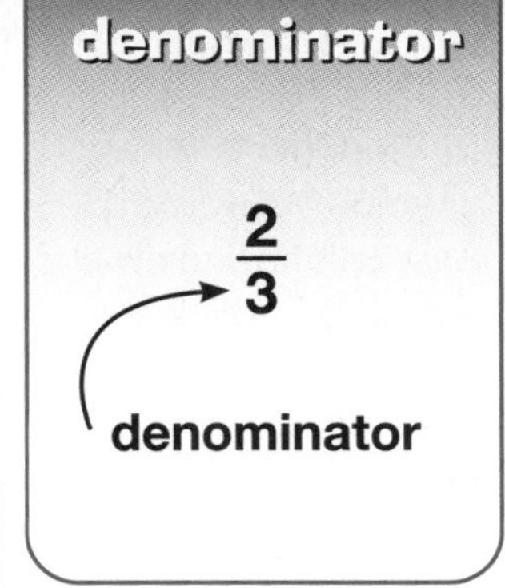

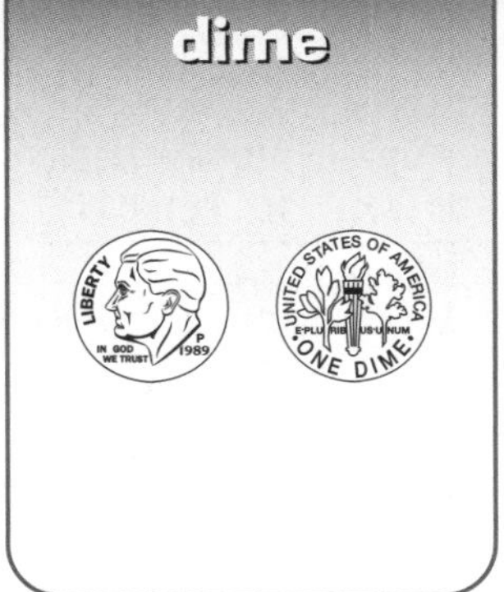

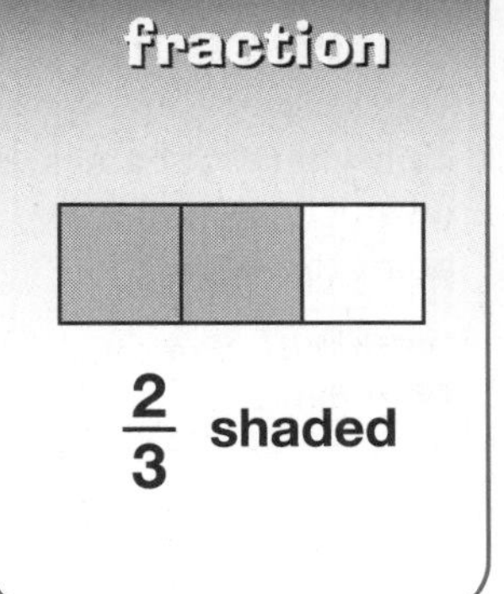

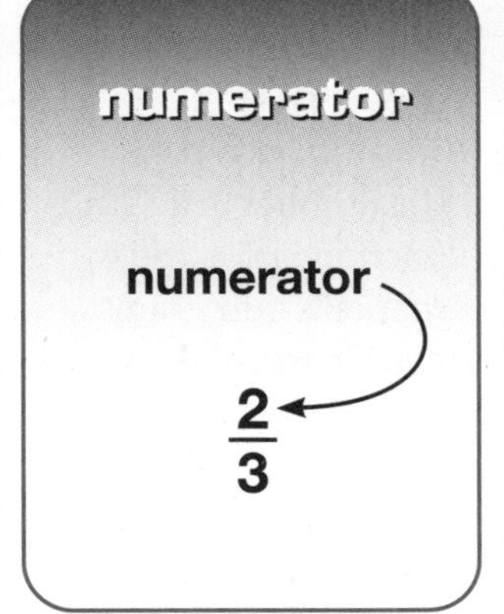

Math Vocabulary

Unit 5

polygon
a closed figure with straight sides

Draw a polygon.

horizontal
going from left to right

Draw a horizontal line.

face
a flat surface of a solid figure

Draw an example of a face of a solid figure.

edge
a line segment where two faces of a solid figure meet

Use *edge* in a sentence.

vertical
going straight up and down

Draw a vertical line.

triangle
a polygon with three sides

Draw two examples of a triangle.

trapezoid
a four-sided polygon with one pair of parallel sides

Draw a trapezoid.

rhombus
a four-sided polygon with all sides of equal length and with opposite sides parallel

Draw a rhombus.

Unit 6

fraction
a number that names part of a whole or part of a set

Write two examples of a fraction.

dime
a unit of money that is worth 10 cents

Use *dime* in a sentence.

denominator
the number in a fraction below the bar that tells how many equal parts there are in a whole or a set

Write a fraction and circle the denominator.

coins
pieces of metal used as money

List 3 different examples of coins.

quarter
a unit of money that is worth 25 cents

Use *quarter* in a sentence.

penny
a unit of money that is worth 1 cent

Use *penny* in a sentence.

numerator
the number in a fraction above the bar that tells how many equal parts of the whole or the set that you are talking about

Write a fraction and circle the numerator.

nickel
a unit of money that is worth 5 cents

Use *nickel* in a sentence.